333 *MORE*
SCIENCE
TRICKS &
EXPERIMENTS

Brown, Robert J.

333 *MORE*

SCIENCE TRICKS & EXPERIMENTS

ROBERT J. BROWN

TAB TAB BOOKS Inc.
BLUE RIDGE SUMMIT PA 17214

Q164
.B71
1984

FIRST EDITION

FIRST PRINTING

Library of Congress Cataloging in Publication Data

Brown, Robert J., 1907-
    333 more science tricks and experiments.

    Includes index.
    1. Science—Experiments.   2. Scientific recreations.
I. Title.   II. Title: Three hundred thirty-three more
science tricks and experiments.
Q164.B8386 1984   507'.8   84-8878
ISBN 0-8306-0835-4
ISBN 0-8306-1835-X (pbk.)

.B71

Line drawings by Frank W. Bolle. Technical assistance by Arthur Wood.

Contents

333 *MORE*
SCIENCE
TRICKS &
EXPERIMENTS

Introduction

This book contains some "Science for You" experiments that have run in newspapers through the L.A. Times Syndicate. More experiments can be found in 333 Science Tricks and Experiments (TAB book No. 1825).

Teachers, parents, and students will find challenging experiments on topics like inertia and momentum, sound, biology, psychology, gravity and centrifugal force, and others. Many experiments are illustrated.

Many of the experiments are thought to have originated with the author. Many others are improvements and simplifications of existing experiments. Some are corrections of errors found in the experiments in other sources. The author has sought to minimize the possibility of error; he has three erudite and meticulous consultants who check everything. Comments and suggestions are always welcome as the author presents more and more experiments through his column.

This volume is dedicated to a man who worked with the project since he was 5 years old. He is my son, Robert J. Brown, Jr. He is a scientist in industry, yet still enjoys our elementary and junior high school tricks and experiments. He originated many of the experiments and improved many others.

It is also dedicated to my daughter, Betty Brown Yow, who began performing science tricks and experiments for me before she was 8 years old. Their assistance and encouragement have been invaluable through the years. Family cooperation could not have been possible without the love and understanding of my wife, Mary T. Brown.

Chapter 1

Intertia & Momentum

MASS, VELOCITY, MOMENTUM

NEEDED: An apple, a sharp butcher knife, a table top.

EXPERIMENT 1: Hold the knife securely on the table, and drop the apple on it from a height of several feet. The apple will cut itself as it hits the blade.

REASON: The momentum of the apple as it falls will carry it past the blade, even though the friction offered by the blade as it cuts the apple will slow it down considerably.

Mass, Velocity, Momentum

Mass times velocity equals momentum. The farther the apple falls, the greater its momentum becomes.

A BRICK TRICK

NEEDED: A brick and a hammer.

EXPERIMENT 1: Hold the brick above the ground. Hit it with the hammer, and it will break without hurting the hand.

EXPERIMENT 2: Try breaking a half of a brick this way. It is much more difficult. The smaller mass of the brick presents less inertia. More of the force of the hammer blow goes through the small object to the hand. The half brick, or "bat," is about 16 times as difficult to break in the hand.

REASON: The muscles of the hand, arm, and shoulder are flexible and descend slightly as the hammer strikes the brick. The inertia of the brick would tend to keep it at rest, but the momentum of the moving hammer would tend to give the brick a high velocity. The result is that the brittle brick breaks by moving faster where the hammer strikes it than anywhere else.

Bumping Bottles

BUMPING BOTTLES

NEEDED: Smooth-sided bottles suspended on strings so they barely touch each other.

EXPERIMENT: Let one bottle swing against the next. The first bottle will stop, and the motion will be transferred to the last bottle in line. This can be done with two or more bottles.

REASON: Laws of conservation of momentum and elasticity are demonstrated here. The bottles are elastic, allowing the momentum to be transferred through them.

Use thick bottles. They are not likely to break, but be careful of glass.

This works better if the bottles can be suspended by two bridles instead of single strings. Two bridles will confine the motion of the bottles so they do not swing in unwanted directions.

Easy Hammering

EASY HAMMERING

NEEDED: Hammer, nails, wood.
EXPERIMENT: Hold the hammer close to its head and try to

drive a nail with it. Then hold it by the handle, farther from the head, and see how much easier it is to drive the nail.

REASON: When the hammer is held by the handle, further from the heavy head, the velocity of the head is greater when it strikes the nail. This is associated with greater kinetic energy of the head, making it possible to transfer more energy (work) to the nail with each blow, while driving it into the wood.

Inertia is a term applicable here. The hammer, in motion, tends to continue in that motion until stopped by the impact with the nail.

The Loose Handle

THE LOOSE HANDLE

NEEDED: A hammer or other implement with a loose handle.

EXPERIMENT: Try to put the handle into the implement tightly by hitting the heavy head itself. Then try hitting the handle. The hammer tightens when the handle is hit in line with its long axis.

REASON: The heavy iron has inertia—more inertia than the lighter wooden handle. As the handle is hit the energy of the blow is transmitted through it to the hammer, but the hammer's inertia prevents it from moving as quickly as the energy in the handle reaches it. So the handle is driven in.

Inertia is that property of matter which manifests itself as a resistance to any change in the motion of a body (McGraw-Hill Encyclopedia). The inertia of a body is proportional to the mass of the body.

A simplified definition is that if the body is still it tends to remain still; if it is in motion it tends to remain in that motion. This is Newton's first law of motion.

Muscular Molecules

MUSCULAR MOLECULES

NEEDED: String, a weight such as thread spool, newspaper, sticky tape or glue.

EXPERIMENT 1: Attach the spool to the string. Cut paper into a strip, and glue it to the spool or upper part of the string. Whirl the weight (out of doors on a calm day) and the spool will pull outward as far as the string will let it reach, while the paper will form an arc, extending itself very little farther than the spool from the hand.

EXPERIMENT 2: Note how easy it is to turn the page of a newspaper by moving the hands quickly. Inertia of the air beside the paper holds it as the turn is made.

REASON: As the paper moves around with the spool and forms an arc, it is bombarded by air molecules from all sides. The pressure of these molecules keeps the paper in the path opened by the spool.

We live in a great and heavy mass of air, but hardly notice it unless the wind blows hard or we move fast.

And where does inertia come in? Inertia is the tendency of a body to stay at rest or continue at the same speed and in a straight line unless an outside force is applied. Here, there is a slight force tending to push the paper strip outward, but little or no other force to move the strip out of its circular path as it slices its way through the air.

Chapter 2

Sound & Other Vibrations

MR. DOPPLER'S EFFECT

NEEDED: While driving on the highway, notice that an automobile horn has a higher pitch as it approaches than when it has passed and is speeding away.

REASON: Sound is made up of waves in the air. When we are

Mr. Doppler's Effect

approaching the horn, our ears receive more of the waves per second than the horn is producing; if we are receding, our ears will receive fewer per second. This is called the "Doppler principle" because of its discovery by Christian Doppler more than a hundred years ago.

AN AIR HORN

NEEDED: A soft drink bottle in a moving automobile.

EXPERIMENT: Hold the bottle at the car window, changing its position. At one point a whistle will be heard coming from the bottle.

REASON: Air flowing across the opening of the bottle makes a whistle just as a sound may be heard by blowing the breath across a bottle neck opening. Air blown into the bottle mouth swirls into a vortex, the action alternately compressing and expanding the air. The compressions and expansions set up waves of air outside the bottle, and these comprise the sound that reaches the ear.

An Air Horn

TUNING THE DRUM

NEEDED: Observation of a large orchestra.

EXPERIMENT: Observe a large orchestra. The story has it that a small boy, watching the tympani player in the orchestra, wanted to know why the tympanist smells the drum occasionally.

REASON: The musician does not bend close to the drum to smell it, but to listen to it as it is lightly struck with a finger. The head, stretched over the drum, is a membrane that is affected by humidity of the air, and it gets out of tune very easily. The player checks it several times usually in one concert. By holding the head close it is possible to hear the minute sound of the drum and tune it while the other instruments of the orchestra continue to play.

VIBRATIONS THROUGH A CONE

NEEDED: A phonograph and a 78 RPM record, a sheet of stationary, a pin, cellophane tape.

Tuning the Drum

Vibrations Through a Cone

EXPERIMENT: Make a cone of the paper, stick the pin into it as shown, hold the pin in the record grooves as the turntable turns, and the music will be heard coming from the cone.

REASON: A phonograph groove is not straight, but is full of waves. As the waves pass against the needle they set the needle to waving, or vibrating, and the vibrations are transferred to the paper. Since the paper is larger than the needle or pin it sets the air to vibrating more than would the pin alone, and the vibrations are heard by the ear.

A 78 RPM record is necessary for this for two reasons. First, the grooves and waves in a 45 RPM or long-play record are too small to register very loudly, and second, the grooves are so small and fragile that an ordinary pin or needle can damage them.

HEARTBEATS THROUGH A LOUDSPEAKER

NEEDED: A strong public address amplifier, a two-inch loudspeaker, shielded cable, plug.

EXPERIMENT: Solder the plug on one end of the cable, making sure the shield is attached to the correct place. Solder the loudspeaker to the other end. Plug it in, turn up the volume, touch

Heartbeats Through a Loudspeaker

the loudspeaker to the bare chest over the heart, and the beating can be heard.

REASON: The loudspeaker consists mainly of a movable diaphragm attached to a coil of wire that can move in the lines of force of a permanent magnet. Each movement induces a small current in the coil, and this current, transmitted to the amplifier, is changed into vibrations in the air which are heard.

It is necessary to put the public address loudspeaker some distance away to prevent feedback. Ten feet should be sufficiently far away.

Almost any school will have a strong amplifier. More volume may be produced by adding a pre-amplifier. This can be built from an electronics store module costing about $2.

VOICE VIBRATIONS

NEEDED: A can, a rubber sheet (may be cut from a balloon), a piece of mirror, a light colored wall or sheet, sunlight.

EXPERIMENT: Tie the rubber over one end of the can (both ends of the can must be cut out), and glue the small piece of mirror on the rubber, not in the middle. Place the can on a post or hold it very still, adjusting it so sunlight will reflect a small spot from the mirror to the wall. Talk into the open end of the can.

11

Voice Vibrations

REASON: The vibrations of the voice transmitted through the air in the can cause the rubber to vibrate. The small motions of the vibrating rubber are amplified by the light reflection from the mirror. The pattern on the wall will suggest the pattern shown on the screen of an oscilloscope, an electronic device that shows pictures of electrical and other phenomena on a screen similar to a television screen.

A large coffee can works well for a deep adult voice. A slim frozen juice can works better for a child's or woman's voice.

MAKE YOUR OWN DRUM

NEEDED: Coffee can, rubber balloons, a funnel, a lighted candle.

EXPERIMENT: Both ends of the can must be cut out. Stretch balloon rubber over both ends, and hold the rubber with bands or strings. Tap on one end and the rubber sheets will vibrate—it is a small drum!

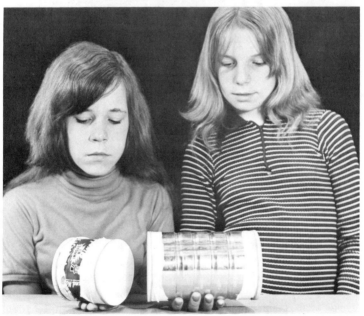

Make Your Own Drum

REASON: The vibrations of the sheet struck with the hand continue through the air in the can, making the other sheet vibrate. If a funnel is held as shown it can concentrate the moving air vibrations so the air from the funnel can snuff out a small candle flame.

The wave travels very fast through the can. A wave traveling through pipe and hose in a long railroad train can reach all the cars in a very short time, setting all the brakes.

Sympathetic Vibrations

SYMPATHETIC VIBRATIONS

NEEDED: A piano.

EXPERIMENT 1: Uncover the strings of the piano by opening the lid. Hold the loud pedal down, releasing the strings so they can vibrate freely. Sing a note, loud, into the piano, and the sound can be heard coming back for a few seconds as a string tuned to the pitch of your voice continues to vibrate.

EXPERIMENT 2: Try other musical instruments to make the piano strings vibrate.

EXPERIMENT 3: If two pianos can be found in the same room, see if one of them can set up sympathetic vibrations in the other.

REASON: Sound is made up of vibrations (in air, in this case) and the vibrations falling on a piano string tuned to the pitch of the voice sound will begin to vibrate "in sympathy." The string will continue to vibrate after the sound that started it cannot be heard.

A Megaphone

A MEGAPHONE

NEEDED: Light cardboard.

EXPERIMENT: Roll the cardboard into megaphone shape. Talk through it to someone several feet away, then talk in the same voice without it. The voice carries farther and is louder when the megaphone is used.

REASON: In ordinary speech the sound waves go out in all directions, getting weaker as they go farther. The megaphone concentrates them so that more of them go in one general direction without losing so much of their energy in that direction.

The Ghost Whistle

THE GHOST WHISTLE

NEEDED: Two whistles such as those used on tea kettles. The whistles must give steady notes. A police type whistle which has a ball inside, giving it an intermittent sound, cannot be used.

REASON: The whistles should have almost the same pitch but not quite. Have two people blow the whistles and a "beat" will be heard. The beat is heard as pulsations which represent the difference in frequency of the two sounds made by the whistles. Have the people blowing the whistles stand at different distances from each other and the effect can be changed somewhat. Try this by having two people whistle with their lips. Many effects can be produced.

BIG SEA, LITTLE SEA

NEEDED: Two jars of different sizes.

Big Sea, Little Sea

EXPERIMENT: Everyone has heard that if a seashell is held to the ear the sounds of the sea may be heard. Here we see that jars serve the same purpose, but a deep heavy "sea" is heard in the big jar, and a light, higher-pitched "sea" sound is heard in the smaller jar.

REASON: The sounds heard are "sympathetic" vibrations, which means that they are sound waves in the room or out of doors that in turn set up waves inside the jar. The air in the jar can vibrate normally at certain pitches, depending on the size and shape of the jar. A little air in a little jar can vibrate at higher pitches than more air in a larger jar.

PENDULUMS

NEEDED: A spring from an old window shade roller, some nails or other weight, some string.

EXPERIMENT: Suspend the spring and the string. Attach weights to each. The weight on the string will swing back and forth in the manner of the pendulum. Its path is not always in one dimension, but the outline of the swing is a long narrow ellipse. The weight on the spring may be made to go up and down or to twist. The outline of its swing may also be an ellipse. This gives us three types of pendulums.

17

Pendulums

WAVES SHOWN BY "SLINKY"

NEEDED: A "Slinky" spring and a flat slick table top.

EXPERIMENT: Hold the spring as shown, flick it with the finger, and a wave will travel through its length. The wave may be seen very distinctly.

REASON: As one coil of the spring is moved, it transfers its energy of movement to the next coil then returns to its former position. The second coil does the same, then the third, etc. until the wave has traveled the length of the spring.

Waves Shown
by "Slinky"

The same thing happens in the air when a sound is produced, except that the wave travels from the source in all directions radially. A small amount of air is moved; it transfers its energy to the air next to it, and so on. Air with sound waves moving outward is in a state of alternating invisible condensations and rarefractions of its molecules.

Chapter 3

Projects to Build

A MEASURE OF DISTANCE

NEEDED: Wood, bolts and nuts, a small spring, paint or crayon, felt or tape, glue.

EXPERIMENT: Build the device as shown in the drawing. Make the wheel 11.46 inches in diameter including a felt edge. The felt reduces slipping while in use.

The author used, as a handle, a 1 × 3 board sawed as shown. Plywood was used as the wheel. A painted mark or circle near the edge of the wheel shows a starting point for the wheel and allows visual counting of the revolutions. A click of the spring against a bolt with each revolution serves for auditory counting. Each revolution

A Measure of Distance

of the wheel signifies one yard distance.

Highway distances are frequently measured with such a device. (Suggested by the Nuffield Junior Science Apparatus book.)

The Glass Harmonica

THE GLASS HARMONICA

NEEDED: An assortment of drinking glasses with stems, wood, screws, felt, water, vinegar.

EXPERIMENT: Hold a glass with the fingers of one hand securely on the table. Dip a finger into vinegar, rub it around the rim, and a musical tone should be produced. Try this with several glasses, adding or taking away water to adjust the pitch. The musical scale should be produced. Mount the glasses in line on a board. They may be mounted with three cleats to a glass. Felt should be glued on the bottom of the cleats to avoid cracking the flanges off the glasses. Or, they may be mounted as shown in the illustration. Here two strips of wood are used to hold all the glasses in the row. Quarter-inch foam rubber is glued on the bottom of the strips to protect the glasses as the strips are tightened down with wood screws.

With nine glasses, all tuned, many beautiful melodies with harmonies may be played if two hands are used, as shown in the photo. Larger glass harmonicas, as used by professionals, have many more glasses, arranged in two rows. Four may be played at a time to produce four-part harmonies.

A BUZZER

NEEDED: Magnet wire, corner braces and screws from a hardware store, wood, tin can metal, a hack saw, soldering equip-

A Buzzer

ment, bolts and nuts, washers.

EXPERIMENT: Put washers on the bolts, put nuts on them, leaving room for them to go through the brace and be held with another nut. Wind magnet wire around them, making sure it is wound in opposite directions on the two bolts. Cut one brace off at the first hole. Put a thumbscrew through that hole. Attach both braces to the wood base as shown at A. First C is a switch made of tin can metal; next C is tin can metal screwed to the wood base so it can vibrate at its upper end. The arrow points to the piece of brace that was cut off—it is soldered to the can metal.

REASON: When the switch is closed both magnets (bolts with wire) attract the can metal and the heavier iron that is soldered to it. When the attraction pulls the metal away from the thumbscrew in the first brace the path of the current is broken, and the magnets allow the can metal and heavy iron to swing back, to make contact again.

The author wound 55 turns of No. 20 magnet wire around each bolt, as at B. The bolts used were quarter inch, two inches long. Braces were 1/8 inch thick, 2 1/2 inches to a side. This buzzer operates well on one D size flashlight cell.

Note that one wire from the magnets goes to the cell, the other to the tin can metal under the magnets. The magnets do not touch this metal.

A WAVE MACHINE

NEEDED: One piece of .048 steel wire 36 inches long, 54 pieces of 1/8 inch steel drill or welding rods, wood, screws, tin can metal, soldering or welding equipment.

EXPERIMENT: Make a form out of wood in which grooves are cut and equally spaced. The rods are placed in this for soldering. They are spaced 5/8 inch apart on centers. Make angles from tin can metal; cut v-grooves in them to support the steel wire on a plywood board. The steel wire must be free to turn in the grooves.

REASON: Move the end rod up and down, and the "wave" will travel to the other end of the assembly, reflect there, and return. Move the end rod up and down regularly, and a standing wave is produced. Attach an Erector motor to the end rod, so that an arm moves the rod up and down regularly, and the standing wave is made without the touch of hands.

A Wave Machine

A MAGNETIZER

NEEDED: A cardboard tube from bathroom tissue, tape, rubber or plastic covered house wire, size 14, green twisted lamp cord (some lamp cord is too small), a male and female plug, switch, electric iron or heater, and soldering equipment.

The author stuffed dowels into the tube, and fitted plywood squares on the dowels to aid in the winding, as shown in photo A.

23

Tape running through the tube with the dowels is used to hold the turns of wire together after the winding is complete. Push the squares against the ends of the tube, and start winding, using the No. 14 wire. Wind about 120 turns of the wire around the tube, tie the tape around it to hold it, then remove the dowels and squares. Put more tape through the coil and tie it, as shown in photo B.

Connect the lamp cord as shown in photo C. A male plug goes on one end of it, the female on the other, and the switch in the center. Be sure to solder and tape the three joints in the wire.

House current can kill. But this magnetizer is perfectly safe if the joints are well soldered and taped. To magnetize a steel object, insert it into the coil as shown in photo D. The electric heating device, connected as shown, draws enough current through the coil to make it a rather strong magnet, and its pull can be felt in the screwdriver or other object held in it.

Turn the switch on *and off* while the screwdriver is in the coil, and it will become magnetized. Test it by picking up paper clips or nails. To demagnetize, put the screwdriver into the coil, turn on the switch, and remove the screwdriver *while the current is on.*

A Magnetizer

A Smaller Magnetizer

A SMALLER MAGNETIZER

A smaller magnetizer-demagnetizer may be operated from the safe current given by a toy train transformer. It can magnetize needles, and demagentize them, as the larger one can handle larger steel objects.

A small non-metallic tube is needed. The author used the barrel of a cartridge type fountain pen, and wound 64 turns of magnet wire around it. Connect one end of the wire to the transformer, and hold the other end in the hand so it may be touched to the other transformer terminal. A switch is not needed.

Alternating current must be used.

REASON: House current is "alternating" current. That means it reverses its direction, and in the United States the reversals are 120 per second. Each time the current flows in one direction in the wires, it magnetizes the tool in the coil in one direction. When the current reverses, the magnetism in the tool is killed, and takes place in the other direction. What was formerly a south pole becomes a north pole, and vice versa.

When the current is turned off while the tool is inside the coil the tool retains some of its magnetism of the moment. There is no way to know which end will be north and which end will be south.

When the tool is withdrawn slowly while the current is on the magnetism induced in the tool becomes weaker and weaker as it is reversed and the tool is moved farther and farther from the inside of the coil. By the time the tool is entirely out of the coil the magnetism remaining in it is so weak that it cannot be easily detected.

Common electric current in the United States is 60 cycle alternating. This means, according to the definition in *The Basic*

Dictionary of Science (Macmillan), that the current changes from its greatest value in one direction to its greatest value in the other direction and back again 60 times per second. The word "cycles" is being replaced by the word "hertz," which is abbreviated "Hz."

ON-SHORE AND OFF-SHORE WINDS

NEEDED: Cardboard box, with pan or dish of water to fill half of it, dark soil, charcoal, or wallboard painted black to fill the other side, incense for smoke, a photoflood or heat lamp, a knife, transparent plastic wrap.

EXPERIMENT: Leave a peep-hole in the side, covered with the plastic sheet. Cover the top with the plastic to keep out room air movements. Have the water slightly warmer than room temperature at the start of the experiment.

REASON: When the incense is lighted, and the box closed, the smoke will move over the water, because the warmer water warms the air above it, and starts the air moving up, then down toward the dark soil. Turn on the light. As the soil becomes warmer than the water, the air movement is reversed. The smoke will blow slowly toward the soil, up, over the other side of the box, then down

On-Shore Off-Shore Winds

over the water. This shows how air currents blow from the ocean to land when the land is warmer than the water.

Make a Light Engine

MAKE A LIGHT ENGINE

NEEDED: Plywood or stiff cardboard, compass for drawing circles, rubber bands, a sewing machine bobbin with holes (from dime store), nail, screws, wood, tools, a heat lamp.

EXPERIMENT: Assemble the engine as shown, so the plywood ring is supported on the bobbin with rubber as spokes. Make it balance. Shine the infra-red light from the lamp on one side of the wheel, and it will turn.

REASON: The heat will expand the rubber bands, throwing the wheel out of balance.

The opposite is true: heat contracts rubber rather than expanding it. The wheel will turn the opposite way. (If the wheel is well balanced, see if bright sunlight will turn it. Place the wheel in the sun, covering half of it so the light shines only on the other half.)

A Trick Light Box

A TRICK LIGHT BOX

NEEDED: A box, two lamps in sockets, two switches, glass, wood, candle, jar, lamp cord and plug.

EXPERIMENT: Build the box with lid. Paint the inside dull black. Arrange the glass and wood as shown. The author's box is 24 inches square by 7 1/2 inches deep, made of quarter-inch plywood.

Look through the peephole, and turn on one switch. The jar shows. Turn it off and turn on the other; the candle shows. Turn on both at the same time, and the candle is seen inside the jar.

REASON: The glass both reflects and transmits light. More light is transmitted when the bright object is beyond the glass. More is reflected when the bright object is on the same side of the glass as is the peep hole.

It is necessary to move the objects while both lights are on to mount them in proper position.

A SIMPLE SHOCKER

NEEDED: A toy train transformer and two rods to hold.

A Simple Shocker

EXPERIMENT: Turn the transformer on low. Squeeze the rods in the hands, and touch them to the transformer terminals. If no shock is felt, turn the transformer to a higher setting.

REASON: The low voltage from the transformer cannot force enough current through the body to be dangerous, since the normal resistance of the human body is about 50,000 ohms. But do not use any other kind of transformer for this.

The author used carbon rods out of dry cells. They give a good contact when held in the hands. Large nails will do—or any kind of metal rods.

A STRONGER SHOCKER

NEEDED: A radio output transformer, Midland 15-611 or equivalent, a flashlight cell, tin can metal, wire, wood for a case (a

A Stronger Shocker

cigar box will do), wire, screws, or large-headed tacks.

EXPERIMENT: Connect one of the small wires to the cell. Connect the two larger wires to the tacks that will be used as terminals for the fingers. Connect the other small wire and the other end of the cell to the push switch which can be made of tin can metal.

Have someone touch the tack terminals lightly. Jiggle the switch up and down rapidly, and a shock is felt at the terminals. No shock is felt if the switch is held closed.

REASON: The transformer steps up the voltage only when the primary current is turned on and off quickly at the switch. For a weaker shocker, use a smaller transformer, such as Olson T-231.

A SAFE SHOCKER

NEEDED: Three dry cells or four flashlight "D" cells, a small bell transformer, a bell or buzzer, a push button, some wires, a board for mounting the device, some furniture gliders or other metal pieces to serve as electrodes.

EXPERIMENT: Mount the parts on the board as shown.

A Safe Shocker

When the bell is ringing, terminals 1 and 2 will give a shock; terminals 3 and 4 will give a much stronger shock which may be felt through a line of 15 people holding hands.

REASON: Voltage from the cells is too low to give a shock, but the coils in the bell and the making and breaking of the circuit by the points in the bell can build up peak momentary voltages above 100. This is enough to give a shock.

The peaks, going through the bell transformer backwards as shown in the diagram, are increased 6-16 times, and this higher voltage is felt from terminals 3 and 4. Two or three cells may be used to increase or decrease the shock.

All of these shockers are safe.

BLINKY THE BLINKER

NEEDED: A battery of 90 or more volts (D), two neon lamps, NE51 or equivalent (A); one condenser of about one microfarad capacity (B); two resistors, 1/4 or 1/2 watt, one megohm each (C); wire for connections and an empty cigar box.

EXPERIMENT: Connect the parts as shown in the diagram. (If only one blinker light is wanted, use diagram at right with one resistor.) The author made a cover for the box out of felt, leaving holes for the lamps to stick through to make the blinking eyes. This makes a good conversation piece, since the blinking continues for months, with no way to switch it off. Several bulbs may be con-

Blinky the Blinker

nected to the same battery, using the diagram at the right, and they will blink without any regular sequence.

REASON: When the circuit is completed, one neon bulb fires before the other (it would be a very rare coincidence if both fired at once.) While the first lamp is glowing the condenser charges through the resistance in series with the second lamp. The polarity of the condenser is such that as it builds up a charge the voltage across the first lamp is decreased to its extinguishing point. Then the second lamp begins to glow and the condenser begins to charge in the opposite direction through the other resistance. The voltage in the glowing lamp begins to decrease, until the lamp is extinguished, and the cycle continues to repeat over and over.

THE HILSCH TUBE MYSTERY

NEEDED: Vacuum cleaner hose that will blow air *out*, cardboard or wood box, two thermometers, a book, paper, glue.

EXPERIMENT: Make a hole in the middle of the box for the hose end. Place thermometers in both ends of the box. Glue the paper on one end, place the book at the other end to partially lock the air flow. Hold the hose at an angle as shown. Place the book so that some air will flow from the paper end of the box; the blowing paper will be the indicator.

Read the thermometers. Let the air blow five minutes, then

The Hilsch Tube Mystery

read again. The temperature shown on the two thermometers will be slightly different.

REASON: Physicists can talk for hours on the principle that produces this effect, yet it is not completely understood. Well-made Hilsch tubes, running on compressed air, produce large differences in temperature. Such tubes are used in suits worn by workers in places contaminated by radioactivity to cool and keep them comfortable.

The author built the simple apparatus shown in the drawing. When the room temperature was 70 degrees he obtained readings of 93 degrees at one end and 99 at the other. The cleaner produced quite a bit of heat, warming the air that came from the hose.

A RHEOSTAT

NEEDED: A battery, flashlight bulb and socket, a long pencil (No. 2 or 3 lead), knife, tin can metal and shears to cut it, screws, tools, wire, wood.

EXPERIMENT: Assemble the rheostat as shown. When the lever is moved to the left along the graphite pencil lead, the light burns brighter.

A Rheostat

A Better Rheostat

REASON: The graphite offers resistance to the current, and the more of its length the current has to travel the more resistance there is. Resistance reduces the amount of current that can flow in the circuit. As is seen, the pencil must be split carefully to expose the graphite so the lever can touch it along its entire length. Do not break it. A six-volt battery and bulb probably will have to be used, since the graphite offers too much resistance for a smaller voltage.

Rheostats are used to control the sound volume in radios and televisions, and for other uses where variable amounts of current are needed.

A BETTER RHEOSTAT

NEEDED: A pen-light cell, a window shade spring, flashlight bulb, wood, screws, tin can metal, tools.

EXPERIMENT: Assemble the parts as shown, and when the lever is moved over the spring, the lamp glows brighter or dimmer.

REASON: The window shade spring is a conductor, but not the best. As the lever is moved so that more of the spring is in the circuit the lamp dims because a greater amount of spring offers a greater amount of resistance to the current.

The author, in building the model, placed a piece of half-inch dowel (from a lumber yard or hardware store) inside the spring to hold it sufficiently rigid.

It is good to clean the upper surface of the spring, where the lever makes contact, with fine sandpaper or steel wool. This will give better contact. The lever is made of tin can metal folded over to make it stiff enough.

Chapter 4

Tricks

A TRICKY PUZZLE

NEEDED: A glass of water, a magnet, tacks, and small nails. Put the tacks and nails into the water.

EXPERIMENT: Challenge someone to get the metal out of the glass of water without pouring out the water and without reaching into the water.

Hold the glass up, and place the magnet opposite the tacks and

A Tricky Puzzle

nails, touching the glass. The metal will be attracted to the magnet, and may be made to slide up the side of the glass. As it reaches the top of the glass it will cling to the magnet and can be moved away.

This cannot be done easily with a tin can used instead of the glass. The magnetic lines go through glass, or aluminum, or any other non-magnetic material, but not so readily through the metal of the tin can. This is because a "tin" can is only about two percent tin. The rest is steel, which is magnetic.

THE RISING BALL

NEEDED: A ball, a glass jar, a table top.

EXPERIMENT: Place the jar over the ball, and move the jar with a rapid rotating motion until the ball leaves the table top and ascends up into the jar. Keep the jar rotating, and it may be lifted from the table with the ball inside.

REASON: Centrifugal force may be defined as a force that tends to make rotating bodies move away from the center of rota-

The Rising Ball

JAR
BALL
TABLE

tion. As the ball moves away from that center it climbs up the rim of the jar and remains there as long as the jar is rotated rapidly in a circle.

The ball must be large enough that its center as it rests on the table is higher than the rim of the glass jar. Then it can climb up the rim. A large light-weight ball such as a tennis ball is good for this experiment.

AN IMPOSSIBLE TRICK

NEEDED: Writing paper and scissors.

EXPERIMENT: Cut the paper as shown, hold it by the outer edges, pull or jerk, trying to make it tear apart at both cuts at the same time. It is impossible in almost every try.

REASON: No matter how carefully the cuts are made, they are never quite the same. The pull at the cuts is never quite the same. And the paper itself cannot be exactly alike at both cuts. For these reasons the paper does not tear alike at both cuts. It is more likely to tear at both cuts when a quick jerk rather than a steady pull is applied to the paper.

An Impossible Trick

The trick can be done sometimes if the middle section of paper is weighted so that its inertia tends to keep it in the same place as the outer sections are torn away.

MAGIC PAPER

NEEDED: Tissue paper, a gummed envelope flap, a dish, a match.

EXPERIMENT: Roll the tissue paper and stick the edges with tiny pieces of the gummed flap of the envelope. Stand it on the dish, set fire to the top, and as the flame burns toward the bottom the remains of the tube will rise into the air.

REASON: The heat of the flame produces an upward current of air—enough to lift the light-weight paper and ash.

The idea for this came from the magicians' magazine *The Linking Ring*. Magician Bill Pitts tells in the magazine how he extracts a mysteriously vanished dollar bill from the burning tube. That is magic; the science part is the rising of the burning tube.

Magic Paper

The paper rolled into a tube may be stuck with tiny bits of rubber cement instead of paper. Try different size rolls and different paper. Magician Pitts recommended a roll two inches in diameter and eight inches high.

As always, be careful of fire. It is always dangerous.

WHICH ONE'S WHICH?

NEEDED: Two people.

EXPERIMENT: Have someone cross his hands and hold out his fingers. Point to one finger and ask him to move it quickly. Often he will move the wrong finger.

REASON: The eyes and the brain do not work together in the normal pattern when the hands are crossed. We respond in a manner to which we have been trained. We can learn new behavior, but it is a slow process requiring considerable practice.

With practice the right finger can be moved when pointed to.

Which One's Which

A Cool Book

A COOL BOOK

NEEDED: A book with a slick-paper dust jacket.

EXPERIMENT: Hold the book as shown, and let it slide quickly through the hands until it comes to rest on the knees. It will feel cool as it comes to rest between the hands.

REASON: This was in the March, 1975, issue of the magazine *The Physics Teacher.* It was discovered by Dr. Robert Everett Vermillion, of the University of North Carolina at Charlotte, and was presented with the invitation to readers to explain it.

It is this author's suggestion that, as the book comes to rest, the hands press more tightly against it, allowing heat to travel more readily from the hands to the book.

PICTURE TRANSFER

NEEDED: Colored comics from the newspaper, turpentine, paper towels, detergent, water, a spoon, a dish, school tablet paper, a jar.

EXPERIMENT: Put one ounce of detergent, two ounces of turpentine, and four ounces of water into the jar, tighten the lid, and shake until an emulsion is formed. It will be white. Dip a newspaper picture in the solution, blot it lightly with a paper towel, then lay it face down on the paper. Rub it thoroughly with the bowl of the spoon, lift it off, and the picture will be seen in reverse on the tablet paper.

Picture Transfer

REASON: Printer's ink is dissolved by the turpentine, and so is transferred to the other paper. The purpose of the detergent is to make the turpentine droplets remain uncombined to form a uniform emulsion.

The Match Box Drop

THE MATCH BOX DROP

NEEDED: An ordinary small box of matches.

EXPERIMENT: Open the box half an inch, and hold it in the hand so a friend cannot see that it is open. Hold it a foot above the table and drop it. It will stand on end. When the friend tries it, it falls over.

REASON: Of course, the friend tries the trick with the box closed.

When it is dropped open the closing of the box as it hits the table slows the downward motion, allowing the box to settle more easily and not bounce.

THE STRAW-THROUGH-POTATO-TRICK (1)

NEEDED: Soda straws, raw potatoes, a needle.

EXPERIMENT: Unless the potatoes are fresh, they should be soaked in water to soften the peel. Hold the straw firmly between thumb and finger, hit the potato quickly with the end of the straw. With a little practice the straw may be made to go all the way through the potato.

REASON: The straw is a very strong structural shape, and, for this reason, if it is straight, it may be forced through the potato. The myth persists that air trapped in the straw makes it strong enough to penetrate the potato.

Now, make needle holes in the straw, being careful not to bend or crush it. It will still penetrate as well, sometimes easier, thus disproving the myth. If, after practice and an hour's soaking of the potatoes, this does not work easily, use another type of soda straw.

The Straw-Through-Potato Trick (2)

ANOTHER STRAW-THROUGH-POTATO TRICK (2)

NEEDED: Straws with a flexible section, tin can metal, tin snips, potatoes.

EXPERIMENT: Cut a piece of the tin can metal so that it slides over the straw through the grooves as shown. Then, by pushing against the metal the straw may be thrust through the raw potato without closing the passage through the straw at all.

REASON: This is another proof that the air trapped in the straw has nothing to do with the experiment. The straw can penetrate the potato simply because it is a tube, and tubes are relatively strong if force is exerted along their length.

INVISIBLE WRITING

NEEDED: Salt, water, writing paper, carbon paper.

EXPERIMENT: Use as much salt in quarter of a glass of water as will dissolve. Let it set until the undissolved crystals have settled. Write with a toothpick, using the salt solution.

When the writing is dry it is invisible, or practically so. But the salt water has left rough crystals on the paper. When the carbon paper is rubbed over the writing paper the writing stands out and can be easily read.

REASON: The salt crystals have rubbed off more of the carbon than the smooth paper.

LIGHT FROM THE TEETH

NEEDED: A totally dark room, a mirror, wintergreen Life Saver candy, pliers.

EXPERIMENT: Stay in a dark room until the eyes are adapted to the darkness (usually ten to fifteen minutes.) Look into the mirror, and, with lips apart, bite through the mint. A flash of bluish light is seen as the candy is crushed.

REASON: Certain solids when crushed or fractured produce a difference of electrical charge on the pieces. Wintergreen oil absorbed on sugar is one of them. The electricity discharges across the surface of the solid to equalize the charges, exciting molecules or atoms of the solid or the air at its surface, thus producing the light. The small amount of electricity produced cannot be felt. But to avoid any possibility of breaking a tooth the mint may be crushed with pliers to produce the light flash.

THE EDIBLE CANDLE

NEEDED: An apple, almond nut slivers, a knife, a match.

EXPERIMENT: Cut a piece of apple to look like a candle. Place a sliver of almond in the top to make a wick. Light it, and it will burn like a candle.

REASON: Oil in the nut meat will burn for a few seconds like a candle wick. The apple does not burn and may be eaten.

A CHEMICAL TRICK

NEEDED: Cornstarch, tincture of iodine, a glass of water, pan for boiling.

EXPERIMENT: Put three drops of the household iodine into the glass of water. It should turn slightly brown. Have a friend stir it with a finger; nothing happens. Now stir it with your finger; and the color changes to blue.

REASON: Prior to the demonstration, boil half a teaspoonful of cornstarch in enough water to fill a glass half-full. Let it cool. Before the demonstration, dip a finger into the starch water. Then when that finger is used to stir the iodine water the familiar test for starch takes place—the blue color appears. Iodine and starch combine to form a somewhat mysterious temporary combination called "starch iodide", which is blue.

WEDGE MAGIC

NEEDED: Vase or detergent jug, soft rope, a rubber ball.

EXPERIMENT: Blackstone, the magician, used a two-foot rope, placing one end of it into the vase, winding the rest around the neck of the vase. He turned the vase over, letting the rope fall free. It did not fall out of the vase. Then, holding the rope, he turned the whole thing over, and held the vase suspended by the rope. Then he handed out both vase and rope for inspection.

REASON: Inside the vase is a small rubber ball. Its diameter

Wedge Magic

plus that of the rope must be slightly greater than the diameter of the neck of the vase. When the vase is upturned, the ball wedges itself against the rope in the neck and holds tightly. Turn the vase upright again, push the rope in slightly to release the ball, pull the rope out, and hand it to the spectator for examination. Take the vase by the neck, and in handing it to the spectator for examination, invert it so the ball falls out into the hand where it can be concealed.

SYMPATHETIC VIBRATIONS

NEEDED: Two drinking glasses (goblets are best), some vinegar, water, a fine wire.

EXPERIMENT: Put water into the glasses, put a little vinegar into each glass, and dip a finger into the vinegar water. Rub the finger around the rim of each glass, making a musical tone. Put in or take out water from the second glass until the pitch is exactly the same in both glasses. Glasses with stems are easier to hold and vibrate better. Place the wire across the rim of the second glass, and as a musical tone is produced in the first glass the wire will be seen to move slightly on the rim of the second.

REASON: The sound from the first glass sets up resonant vibrations in the second because the two glasses have the same frequency of vibration. This idea is used in musical instruments.

If good thin glasses are used and they will vibrate well, try producing a tone in the first glass, then putting a hand over it to stop the sound quickly. The sound should then be heard coming from the second glass.

A GRAVITY TRICK

NEEDED: Thin plywood or thick cardboard, a knife, a leather belt.

EXPERIMENT: Cut the plywood in the shape shown. When the belt is placed in the slot, the wood with the belt can be held on the finger, seeming at first glance to defy gravity.

REASON: The belt pulls down on the piece of wood. As the large end of the wood starts to move down, the ends of the belt move closer to the body than the supporting finger. The belt and wood may be held at rest in this position because the center of gravity is directly under the finger. This trick will not work with an ordinary weight, but requires a leather belt which is stiff enough not to bend easily.

A Gravity Trick

BERNOULLI WITH A DIME

NEEDED: A dime and a smooth table top.

EXPERIMENT: Challenge a friend to place the dime on the table, then without using his hands or arms, turn it over.

REASON: Blow hard over the top of the coin. Bernoulli learned that air in motion has less lateral pressure than air at rest. The air moving over the coin has less pressure than the air below it, and so the air below flips it over. The coin cannot be turned over by blowing air down on it.

A BRIDGE OF KNIVES

NEEDED: Four table knives and five objects, (cups will do).

EXPERIMENT: Build the bridge of knives as shown, and the fifth cup can be supported on the knife blades.

REASON: Each knife is supported at both its ends, although to some it would not seem so at first glance. When the extra weight is added each knife bends downward slightly until the strain is proportional to the stress. The upward and downward forces are then equal, and friction prevents the knives from sliding sideways.

A Bridge of Knives

BLOW THE COIN OVER

NEEDED: A funnel and a coin.

Blow the Coin Over

EXPERIMENT: Hold the funnel upright, with the coin inside. Try to turn the coin over by blowing hard into the funnel as shown. It is difficult if not impossible. Cover the spout with a finger and the coin may be blown over.

REASON: With the spout uncovered, some of the air passes through the small spaces between the coin and the side of the funnel, tending to equalize the pressure below and above the coin. When the spout is covered air cannot get out below the coin, and more is blown across the top of the coin. According to the Bernoulli principle, the moving air above the coin has less pressure than the still air below, and tends to lift the coin slightly. Then the air catching below it flips it over.

A SALT TRICK

NEEDED: Soft blotter paper, salt, water, matches.

EXPERIMENT: Heat the water, and dissolve in it all the salt it will take. Soak some strips of blotter paper in the solution, dry them, then light one and let it burn half way. After it cools place matches on it as shown and see how many it will hold. An untreated blotter burned the same way falls into ashes and will hold no weight.

REASON: The heat of the burning paper fuses or melts some of the salt crystals together. When they cool they provide the

A Salt Trick

strength to hold the matches. The author, in trying this, succeeded in placing 44 kitchen matches on a two-inch strip of burned blotter before it broke under the weight.

A FORK AND SPOON TRICK

NEEDED: A glass, a fork, a spoon, two matches.

EXPERIMENT 1: Balance the fork, spoon and match as shown. Set fire to the ends of the match. The wood will burn to the edges of the fork and glass, but the fork and spoon will not fall.

REASON: The wood on the match must be supplied with enough heat to make it burn. When the flame reaches the glass and the fork, they conduct so much of the heat of the flame away that the wood does not burn farther. Enough of the solid wood remains to hold the weight.

A Fork and Spoon Trick

EXPERIMENT 2: Lay a match and a cigarette side by side on a piece of board. Light both. The wooden match will not burn past the edge of the wood; the cigarette will burn all the way. The cigarette contains a chemical that promotes burning, to prevent its going out while in the hand. This is good for the smoker, but makes the cigarette more dangerous as a fire hazard.

Chapter 5

Biology & Psychology

EGG CANDLING

NEEDED: A cardboard tube or box, an electric lamp, a dark room.

EXPERIMENT: Cut a hole just smaller than the size of an egg in the cardboard, and mount the light some way behind the hole. Hold an egg to the hole, broad end upward, and look carefully.

COMMENT: If the egg contents do not fill the shell the egg is not perfectly fresh. The larger the air space the older the egg. The yolk should be perfectly clear and round in outline. If, besides the air space, there is a dark haze or cloud in the egg, it has spoiled. Any egg from a store will show air space; this does not mean it is not perfectly good for eating. (Suggested by *Organic Gardening and Farming,* April, 1974.)

LAYERING

NEEDED: Rose bush, ivy, or other plants.

EXPERIMENT: Bend the plant down, cover a section with moist earth, and that section will take root and form a new plant.

COMMENT: Many plants may be propagated in this way. The process is called "layering." Strawberries, ivy, and Virginia creeper, gooseberries, blackberries, and forsythia carry out the layering process without the help of the gardener.

A grape vine may be bent down, run along in a trench in the ground, and new shoots will rise. Covered with damp soil, the

Egg Candling

Layering

shoots will form roots, and the shoots with roots may be cut away. They are new plants, and the parent vine is not injured.

THINK WARM

NEEDED: A small thermometer.

EXPERIMENT: When the hands are cold, hold the thermometer covered by the hands and think "My hands are getting warmer." See if they get warmer. If the hands are warm, see if, by repeating the suggestion that they are getting cooler, they do get cooler. Repeat the suggestions over and over.

COMMENT: Dr. Edward Taub, psychologist at the Institute for Behavioral Research, Silver Spring, Maryland, taught 19 out of 20 subjects to change their skin temperature by using what he called "thermal imagery" and biofeedback.

Without technical aids, 10-year-old Beth Klinstiver, a fifth grade pupil in Eaton's School near Lenoir City, Tennessee, was able to change the temperature of her hands three degrees by "thinking warm" alone. She first held the thermometer tightly in her hands 15 minutes to bring the reading to the normal temperature of her hands.

This is a safe psychological experiment for boys and girls to try. (Suggested by an article in *Science News* October 26, 1974.)

Think Warm

A Cucumber in a Bottle

A CUCUMBER IN A BOTTLE

NEEDED: A growing plant, a bottle, patience.

EXPERIMENT 1: Put the cucumber plant into the bottle while it is still small enough to go in. Let it grow inside the bottle. It will be a curiosity to those who may not know how it was done.

EXPERIMENT 2: Try this with a squash, pumpkin, or gourd plant. Either of these should fill a larger bottle or jug. If one dies after being inserted into the bottle, try another.

COMMENT: If an entire plant is put into a small-mouth bottle it probably will die. The growing cucumber does not have to be in the open; its food is supplied through its stem from the plant which is growing normally in earth and air.

(Suggested by Bob Sherman in *Nature and Science*.)

GREEN LEAVES IN WINTER

NEEDED: Observation.

EXPERIMENT: Watch someone's lettuce rows or grow some lettuce in a pot out of doors.

COMMENT: Some leaves die and fall at the first sign of cold weather. Others freeze as the temperature drops. Some stay green all winter.

REASON: Just why certain varieties of plants tolerate a loss of

Green Leaves in Winter

heat is not always known. In general, an increase in concentration of cell sap lowers the freezing point of the cells. Some plants have less water (cell sap) than others, and when freezing does take place sharp ice crystals do not damage cell walls as much as in other species. Heat of oxidation is liberated more rapidly in some plants than in others at low temperatures and this offsets winter temperatures to some degree. But if temperature drops enough, nearly all growing things exposed to it will die.

The photo shows children in the author's vegetable garden, where all plants had been killed by cold weather except lettuce.

FALLING RAIN OR SNOW

NEEDED: Observation.

EXPERIMENT 1: While driving along in gently falling rain or snow, notice that the drops or flakes seem to be coming toward you. When the car is stopped, they will be seen to fall straight down.

REASON: The appearance of the drops or flakes coming toward you is due to the fact that you are moving toward them, and the motion is relative. You are moving with the car, and while the drops or flakes may be falling straight down, you are getting closer to them as they fall.

Falling Rain or Snow

EXPERIMENT 2: Often the expression "the rain came down in sheets" is heard. This may be observed sometimes when there is strong wind and a heavy downpour at the same time. The drops are close together as they fall. The wind breaks them into smaller droplets, and together they look almost like solid sheets of water. Watch for this!

EXPERIMENT 3: Drive down a straight road across which snow is drifting. A strong wind seems to be blowing the car in the direction in which the snow is drifting, and you feel that your muscles must fight against a real force.

FALLING LEAVES

NEEDED: Observation.

OBSERVATION: A thin stem called the petiole connects the leaf to the main stem of the plant. Where the connection is made there are special different cells called abscission cells. While the leaf is growing and healthy a chemical produced in the leaf, indoleacetic acid or auxin prevents the abscission cells from forming into a cork-like mass. When the summer is over this mass forms, separating the growing parts of the plant and letting the leaves fall. A scar may be seen where a petiole breaks off, and this scar is made up of abscission cells.

Falling Leaves

BALANCING

EXPERIMENT: 1 Stand on one foot. It is easy. Stand on both feet, close the eyes, then lift one foot. It is almost impossible for most people to stand on one foot with the eyes closed, for more than a few seconds.

EXPERIMENT 2: Try running with the eyes closed. It may be difficult. Be prepared to fall down! A soft grassy place is best for trying this.

REASON: We have balancing organs in our ears. The three semicircular canals are looped tubes placed at right angles to each other and filled with liquid. The tubes are lined with nerve cells connected to the brain. Also, tiny loose particles touch nerve endings when the head is tilted.

In addition to these specialized organs, we depend on other senses to keep our balance. Our eyes are important for this; also, our muscles detect any swaying movement and send messages to the brain to correct it.

Keeping our balance with eyes closed is easier after practice. Also, some people find it easier than others.

COMMENT: One theory of sea sickness is that the various balancing signals we use, those from the ears, from the muscles of

Balancing

the legs and trunk, from visual clues (we *know* that the corners of a room are vertical) in the unusual experience of being tossed around in a ship, give conflicting and contradictory messages. The unaccustomed effort of having to decide which signals to believe and act on results in tension, which results in sickness, malaise.

HOLDING THE BREATH

NEEDED: Observation.

EXPERIMENT: Hold the breath as long as possible. It is painful. A time comes, and very soon, when holding the breath is no longer possible.

REASON: The reasons are not fully understood. But two little glands, called the "carotid bodies" are thought to be the parts of the body that react to breath-holding and send out the distress signals.

The carotid bodies are sense organs about five millimeters long in a man. They are located largely behind the carotid artery in the upper part of the neck. They are sensitive to reduced levels of oxygen in the blood. They are responsible for the increased rate of breathing at high altitudes where the air is thin.

A group of scientists in Harbor General Hospital in Torrance, California, is studying the problem, according to the magazine *Scientific American,* June, 1974.

Holding the Breath

Spallanzani

SPALLANZANI

NEEDED: Two jars with tight lids, boiling equipment, juices.

EXPERIMENT: Place juice in two jars. Boil one as in canning; leave the other cold. Seal the boiling jar while hot. Set both jars aside; the unboiled juice will sour or mold or both.

COMMENT: Lazzaro Spallanzani, Italian biologist who lived 1729 to 1799, showed in this way that air carries microscopic life and that boiling kills the bacteria. That was one of the great discoveries of science.

Spallanzani was the first person to watch bacterial cells divide. He discovered that bats do not need to see in order to dodge wires, although he did not solve the mystery. He discovered that some animals, including salamanders, can grow new legs to replace any that are lost.

He was a teacher at the University of Padua.

REMOVE WARTS

NEEDED: A pencil with rubber eraser.

EXPERIMENT: Tell a little brother or sister "I'll make your wart go away. Each time I touch the wart with the eraser you must say Wart, you must leave me." Then touch the wart lightly and slowly six times with the rubber. The wart may go away overnight.

REASON: This is a psychosomatic cure that is not understood, but doctors know it does work in many cases. Any objects may be used to so "charm" the wart away, a coin may be rubbed on it, for example. The charm works less frequently on older people who may be very skeptical.

If warts cannot be charmed away, a doctor can touch them with a little mild acid, or "burn" them off with high frequency electricity that does not shock. The doctors' methods always work.

One common folklore wart remedy is to steal a neighbor's dishrag, rub it across the warts, and bury the rag in the woods. Another involves a dead cat in a graveyard at midnight. Another involves stump water, water gathered in a hollow stump. You immerse your hand with the warts and say, "Stump water, stump water, take away warts."

Warts are now known to be caused by (communicable) viruses. Most adults have become immune. Hence, warts are usually a childhood condition.

Nobody knows how belief in a particular applied magical cure can do anything to a virus, but it is well-established that it can.

Remove Warts

Grow a Pine Tree

GROW A PINE TREE

NEEDED: A pine cone with seeds, a flower pot, and soil.

EXPERIMENT: Plant a seed from the pine cone, twice as deep as the diameter of the seed. Keep the dirt moist until the seed sprouts. Then place it in direct sunlight and water it about once a week.

COMMENT: Suzanne Brown, the author's granddaughter, found such a cone, took a seed out with a knife, and planted it. It grew. This is a rather unusual experiment for a boy or girl, growing a tree from a pine seed.

Cone-bearing trees, or "conifers" are interesting to study. An encyclopedia has a good section on them. There are two kinds of cones on the same tree, pollen cones and seed cones. The seeds are not enclosed, but are free to fall out from between the scales of the cones. They fall to the ground and some of them grow into trees.

Cones can be as large as 15 inches, or can be very small, depending on the type of tree or shrub.

HOW FAST ARE YOUR SIGNALS?

NEEDED: A friend.

EXPERIMENT: Have someone inflict slight pain, and see

How Fast Are Your Signals?

how quickly the signal travels through nerve fibers to the brain.

COMMENT: All impulses travel so quickly that a boy or girl could not determine which kind travel fastest. But Henri Busignies, writing in "Scientific American," says that each bundle of nerve fibers has its own characteristic conduction speed, from a few meters to about 100 meters per second.

Signals conveying muscle position travel at the highest velocity, he says, presumably because balance and quick movement are vital. Pain signals are among the slowest.

A pinch is a harmless experimental pain.

STERILIZE SOIL

NEEDED: Soil to be sterilized, baking bags, an oven.

EXPERIMENT: Place the soil in the bags, seal them, and heat in the oven an hour or longer at a 250 degree temperature.

REASON: Soil for potting may contain a harmful fungus, or a parasite, that can attack the plants as they grow. Sterilizing at this temperature kills any living thing in the soil. Of course, the soil must cool before a plant is placed in it.

Nematodes, fungi, and weed seeds are killed by sterilizing. A too-high temperature is not desirable, however, because it can destroy some of the nutrients of the soil.

Sterilize Soil

Watch the Roots

WATCH THE ROOTS

NEEDED: Glass jar, white blotting paper, cotton or cotton rags, seeds, water.

EXPERIMENT: Cut the blotter so it fits around the inside of the jar. Place it in the jar, and hold it in place by stuffing cotton or rags loosely behind it. Use a pencil or knitting needle to open space for placing the seeds inside and arranging them, as shown. Keep the blotter moist.

OBSERVATION: The seeds will sprout and the roots and stems may be seen clearly as they grow. By turning the jar upside down the roots and stems will be seen to reverse their direction of growth.

PLANTS AND TEMPERATURE

NEEDED: Two boxes the same size, topsoil, grass sod, two thermometers.

EXPERIMENT: Place soil in one box with a thermometer, place soil with sod above it in the other box with a thermometer. Place both boxes in the sun. The temperature will be higher in the box without the growing sod.

Plants and Temperature

REASON: The leaves of grass—and other green leaves—keep cool by bringing up water from the roots and having it evaporate from the leaves. The evaporating water cools the air.

This explains partially why wooded land is cooler than open fields or deserts. Another reason is that the shade from trees keeps the sunlight from hitting the ground and turning it into heat.

When the author tried this experiment the temperature settled at 89 degrees in the bare soil and 82 degrees in the sod after five hours in the sun.

The bare soil gets colder quicker at night.

TONGUE FOOLERS

NEEDED: Grapefruit juice, water, cocoa or candy, salted peanuts.

EXPERIMENT: Drink a sip of water, then some grapefruit juice. The juice is delicious. Take a sip of cocoa or eat a small piece of candy, then taste the juice. It is very sour. Eat some peanuts or other salty food, and taste the juice again. It is again delicious.

REASON: The juice is acid, and therefore sour, but only mildly sour—and pleasant. Immediately following candy or sugar the same juice seems to be unpleasantly sour by comparison. But

Tongue Foolers

the immediately previous sweet taste is offset by more grapefruit juice or salty food.

If you eat artichokes, Cynara scolymus, all things will taste sweet afterward, even water.

There is an African fruit called miracle fruit that has this property in very high degree. If you eat this fruit, afterward even a lemon will taste as sweet as a tangerine, and quite similar.

AFTERIMAGE, AN EYE TRICK

NEEDED: Black paper, white paper, scissors, paste, a strong light.

EXPERIMENT: Cut a disc two or three inches in diameter from the white paper. Cut a square twice as large from the black paper. Paste the white disc on the middle of the black paper.

Place the papers in strong light and gaze at the white disc for one minute. Then look at a blank wall that is not brightly lighted. The eyes will see the disc again, but this time it will be black.

REASON: When the eyes are exposed to bright light for several seconds the retina shows tiring or fatigue of those retinal cones sensitive to the color observed. When the same area of the retina observes another surface such as the wall dimly lighted the

Afterimage, an Eye Trick

other retinal cones respond more strongly and the complementary color is seen.

"Afterimages" of this kind cause black to become white, white to become black, yellow to become blue.

DON'T BLINK

NEEDED: Two people.

EXPERIMENT 1: Puff breath at the top of a nose, and dare the person to stare and not blink. It is almost impossible to refrain from blinking when the air hits the eyes.

EXPERIMENT 2: Hold a thin sheet of plastic in both hands, and bring it quickly toward a friend's face. The friend knows it could not do harm, yet he will blink. This is a protective reaction as explained below.

Don't Blink

EYE GAMES

NEEDED: A table, two objects, a friend.

EXPERIMENT 1: Have the friend kneel so his eyes will be on a level with the table top. Place one object in the middle of the table, so it does not make a shadow. Have the friend hold a hand over one eye. Place the other object a few inches closer or farther from the friend, and have him guess which it is.

EXPERIMENT 2: Try this with the objects twice as far away from the eye. Also, try it with small objects closer to the eye.

REASON: Our eyes turn slightly inward when they focus on an object, and the brain interprets the turn to show us the distance of the object. This is called stereoscopic vision.

When only one eye is used there is no stereoscopic vision, and it is difficult to tell distances. In this experiment it will be difficult for the friend to tell whether the second object is closer or farther than the first.

Eye Games

COMMENT: Nearly all predatory mammals have excellent stereoscopic vision, their eyes in the front, looking straight ahead. Most of the day-hunting raptorial birds such as hawks and eagles have excellent stereoscopic vision.

Non-predator mammals, such as rabbits and field mice have their eyes on the sides of the head. A rabbit has almost 360° vision in all directions. The only blind spot in the rabbit's vision is a small cone containing his ears.

STARCHY CRACKERS

NEEDED: Crackers, household tincture of iodine, water, a dish.

EXPERIMENT 1: Add two drops of the iodine to a half glass of water. Wet one cracker, and place it on the plate. Chew the other cracker until it tastes sweet, and place it from the mouth to the plate. Put a few drops of the water on each cracker.

EXPERIMENT 2: Chew a piece of highly flavored chewing gum, take it out of the mouth, then chew the cracker. Is there a difference?

EXPERIMENT 3: Start chewing the cracker with a mouthful of soft drink. Is there a difference?

OBSERVATION: The iodine solution placed on the wet cracker turns it a purplish color; that placed on the chewed cracker does not change its color.

REASON: The purple color is a test for starch, showing that

Starchy Crackers

the cracker contains starch. Saliva has changed the starch in the other cracker to sugar, so that there is no color change. Digestion of starch begins in the mouth. And continues in the stomach.

HYDROTROPISM

NEEDED: A cardboard box, screen wire, soil, lima bean seed, a small dish for water.

EXPERIMENT: Make a hole in the bottom of the box and place the screen wire over it. Put a half inch of dirt in the box, plant one or two seeds in it, and cover with another half inch of dirt. The soil should be kept moist. Place the box over the dry dish or saucer.

The roots will start growing downward in the normal way, responding to the pull of gravity. This is called "geotropism." But when they have grown through the screen they turn to the side, seeking water. This seeking of water by the roots is called "hydro-tropism."

REASON: The attraction of roots toward water is stronger than their attraction downward.

The experiment has been performed without water below the box. Now try it again, this time placing a dish of water under the box, so the water comes to within a quarter of an inch of the screen. The roots will grow downward into the water.

Hydrotropism

GEOTROPISM

NEEDED: Milk cartons, dirt, bean or mustard seed.
EXPERIMENT: Cut off the cartons to make flower pots, plant

Geotropism

the seeds in them. Let the seeds sprout until the plants are an inch or two high, then turn some of the cartons on their sides, leaving the others upright.

The earth's gravity causes the roots to grow downward and the stems upward in each case. The root growth may be seen if the cartons are cut away from the dirt.

In the lower drawing, the plant is growing upward from the overturned pot. In the upper drawing the plant has righted itself after the pot was turned upright again.

PLANT FORCE

NEEDED: Pots, dirt, lima bean seeds, six quarters or other weights.

EXPERIMENT: Plant the seeds, one to a pot. Tape two quarters together, and three quarters. Place one quarter over one seed, two taped together over another seed, and three over the other. The force of the growing plants will push all the weights aside.

REASON: The division of cells that we call growth can exert tremendous force. Growing roots can break rocks apart. This force is developed when cells in growing tissues split and enlarge as water and nutrients are absorbed and used to make new cellular materials. "Osmotic turgescence" is the term frequently used to indicate the chief forces involved. (The pots are made from the bottoms of milk cartons.)

PHOTOTROPISM

NEEDED: Milk cartons, soil, bean or mustard seeds.

EXPERIMENT: Cut off the cartons, place dirt in them, and plant the seeds. Place some cartons in a dark place, some where bright light reaches them from one side and others where light is about equal on all sides. The plants will grow toward the brighter light.

REASON: Cells on the darker side of the plant grow faster and longer. The seeds planted in a dark place will grow faster, but will die in a few days for lack of food. Growth regulating substances called "auxins" are produced in the tips of young growing stems, and migrate in greater quantity toward the darker side of the stems.

A TRICK OF THE EYE

NEEDED: A card, a pin, an electric bulb.

EXPERIMENT: Make a pin hole in the card and look at the

A Trick of the Eye

light through it. Bring the pin near the eye and into the line of sight between the hole and the eye. The pin will appear upside down.

REASON: The simplified diagram shows the paths taken by the light rays from the bulb to the eye. The pin in this case is a shadow cast on the retina of the eye. Since the shadow is right side up, the brain will see it as being upside down.

This may not be easy to see on the first try. Hold the pin close to the eye. Use a bright frosted bulb. Hold the card at arm's length.

ANOTHER TRICK OF THE EYE

NEEDED: Cardboard and a heavy needle or pin.

EXPERIMENT: Make a hole in the card with the needle, then have someone who reads with glasses remove the glasses, close one eye and read. Then have him hold the card close to the eye, looking through the hole to read. The printing will be much clearer.

REASON: Because of eye defects the light rays entering the eye, and shown in dotted lines, do not focus on the back of the eye, the retina. The eyeglass lens corrects the fault so the rays all focus correctly. But if the only light entering the eye passes through the center of the cornea or lens it does not have to be focused. It goes straight through to the retina and produces the image, as shown in the heavy line.

Another Trick of the Eye

THE GREAT REDI EXPERIMENT

NEEDED: Three wide mouth jars, three pieces of raw meat, paper, cheesecloth.

EXPERIMENT Place a piece of meat in each jar; cover one with paper, cover one with two thicknesses of cheesecloth, and leave the other uncovered. The meat in all three will putrefy but maggots will appear only in the uncovered jar. (Leave the jars out of doors in warm weather for this experiment.)

REASON: Aristotle taught that maggots appeared spontaneously in putrefying flesh or filth. Franscisco Redi, in 1668, performed this experiment (in a much more elaborate fashion) to prove that life can come only from other life, in this case, maggots can come only from fly eggs.

It may be necessary to protect the jars from nightprowling cats. The author used screen wire with large mesh for this. The mesh must be large enough to allow flies to get through it.

THE PRESSURE OF SWELLING SEEDS

NEEDED: Lima bean seeds, dry sand, water, a pint jar.

EXPERIMENT: Mix half beans and half sand in the jar, shaking the jar and pushing the sand in tightly. Wet the sand, but do not put in enough water to flood it. Screw the lid on tightly (it does not have to be air tight, however.)

The beans will absorb water from the spaces between the sand grains, swell and in a few hours burst the jar. Be sure the jar is placed on a large cookie sheet or other sheet to catch the broken glass and sand.

An Unusual Touch Sensation

AN UNUSUAL TOUCH SENSATION

NEEDED: Two people.

EXPERIMENT: Hold the fingers as shown and rub them with the finger and thumb. The feeling will be a totally unfamiliar one.

REASON: For years we have grown accustomed to touching our bodies in various ways and getting the same sensations each time. If the other person's finger is inserted in a place where we are accustomed to feeling our own finger the difference in the pattern of feeling registered in our brain is quite astonishing. We are all creatures of habit, and one of our important habits is our consistent and dependable sense of touch. In this experiment we are playing a trick on our sense of touch.

SEE ROOT HAIRS

NEEDED: A saucer, a dish, paper towel, radish seeds, water.

EXPERIMENT: Cut a piece of paper towel the diameter of the saucer, put it in the dish, and wet it. Put radish seeds on the wet paper, half an inch apart, invert the saucer over the paper, and leave

See Root Hairs

it for two days in a warm place. The white cotton-like growth seen is made up of many root hairs.

A root hair is a long narrow outward extension of a single surface cell. The root hairs extend into the soil between earth particles, to take up moisture and nutrients for the growth of the plant. A magnifying glass will show them more clearly.

INFECTION

NEEDED: Two good apples, one apple with a rotten spot, needles, matches, string.

EXPERIMENT: Tie the string around the stem of one healthy apple to identify it. Sterilize both needles by heating them in the flame of a match. Let them cool. Stick one needle into the rotten spot, then without cleaning it, stick it into the apple with the string marker. Stick the other needle into the other sound apple, without first sticking it into the rotten spot. Remove both needles, throw away the rotten apple, and place the other in a warm room.

If other factors are equal, the apple with the string should develop a rotten spot where the needle penetrated it.

REASON: Bacteria will have been transferred from the rotting apple to the good one.

PUFFBALLS

NEEDED: A puffball, found in woods or on the lawn.

Puffballs

EXPERIMENT: Kick the puffball lightly with the foot. "Smoke" puffs up from it.

REASON: That "smoke" consists of millions of spores of that particular mushroom plant. They are carried by the breeze, fall to the ground, and are likely to produce other plants in other locations.

Some puffballs may be three feet in diameter. Mushrooms (and puffballs) are fungi, plants that do not produce their food but live on food taken from decaying matter or sometimes from live hosts. "Devil's Snuffbox" is the name sometimes given to the puffball.

A TEST FOR VITAMIN C

NEEDED: Half a teaspoonful of cornstarch, water, tincture of iodine, foods to be tested, (perhaps lemon and orange juice).

EXPERIMENT: Boil the starch in half a glass of water. Put 20 drops of the mixture into a glass of water, and add one or two drops of iodine. A blue color should appear. If food containing vitamin C is added, (drop by drop) the blue color should disappear.

REASON: Food containing vitamin C seems to destroy the somewhat mysterious combination of starch and iodine which is responsible for the blue color. But the color is also destroyed in other ways, so this is not a specific and sure test for vitamin C, in spite of some claims.

Yeast and Carbon Dioxide

YEAST AND CARBON DIOXIDE

NEEDED: Two large jars, small glass, molasses, yeast, lime.

EXPERIMENT: Put limewater into the glass and set it in the large jar. Put a small amount of molasses and yeast in the jar, and put in water until it comes almost to the top of the glass. Screw the lid on the jar, let it set in a warm dark place, and in two days the limewater should be milky rather than clear—a test for carbon dioxide.

REASON: Yeast cells contain enzymes which change the sugar of the molasses to form grain alcohol and carbon dioxide. The carbon dioxide unites with the limewater, producing a small amount of calcium carbonate which stays in suspension, forming the white or milky color. Chemically the limewater test is: $Ca(OH) + CO_2$ $CaCO_3 + H_2O$.

Limewater may be bought in the drug store or made by mixing lime and water, letting the mixture settle until clear liquid may be poured off the top. As a double-check or "control" put a glass of limewater into an empty closed jar at the same time. The limewater will remain clear.

This lime water test can be tricky. If too much carbon dioxide goes into the solution the calcium carbonate becomes calcium bicarbonate, which may dissolve and leave the solution clear again.

Make a Killer Jar

MAKE A KILLER JAR

NEEDED: A jar with a lid, a small piece of screen wire, cotton, some nail polish remover, a small insect.

EXPERIMENT: Place the cotton in the bottom of the jar. Cut the screen and bend it so it will fit down around the inside of the jar, over the bottom. Soak the cotton with nail polish remover and drop the insect into the jar. Close the lid tightly. If the bug does not die in a few minutes, add some more remover. This way of killing insects lets them die relaxed, and their appearance is unharmed.

REASON: All animals and insects require oxygen. Here the oxygen supply is replaced in the air with polish remover fumes which are poisonous in dense concentrations.

EARTHWORMS

NEEDED: A jar, moist earth, sand, cornmeal, dark paper or cardboard.

EXPERIMENT: Dig some moist earth from a place where earthworms are found. Fill the jar about two-thirds full of soil, sprinkle some cornmeal on it, then cover with half an inch of moist sand. Wrap the paper around the jar so no light enters. Keep the top open so the earthworms can get air.

After several days the burrowing worms will have brought earth to the top of the sand. Some of their burrows may be seen against the inside of the glass. Earthworms are harmless and very beneficial in the soil. They continually mix and enrich it, and loosen it so that air may circulate through it.

THE POWER OF SUGGESTION

NEEDED: Two people.

EXPERIMENT: Do not let the other person know that an experiment is being conducted. Scratch your head with a finger, and chances are the other person will do the same. Cough, and the other person is likely to clear his throat or cough.

REASON: The power of suggestion is very strong. We are all subject to it. When the other person sees us scratch, he immediately thinks he needs to do the same. When we cough, we merely suggest it to the other person, and he subconciously acts on the suggestion.

Capture Spider Webs

CAPTURE SPIDER WEBS

NEEDED: A duster, black paper, sifted flour.

EXPERIMENT: Place flour in the duster. Find the spider web in early morning when dew is on it. Dust some flour on it so all the strands are covered lightly. Bring the black paper up behind the web, and lift the web gently on the paper. It will stick.

To make the duster punch a dozen holes in the bottom of an old plastic bottle. The flour is dusted out as the bottle is squeezed.

Chapter 6

Water & Surface Tension

MYSTERY OF THE MELTING ICE

NEEDED: A glass of water and one or two large ice cubes.

EXPERIMENT: Place the ice in the glass, then fill the glass with water just to the overflowing point. Since the ice extends above the water surface, an overflow is expected as it melts. This does not occur.

Mystery of the Melting Ice

REASON: Ice is one of the substances that expands when freezing. It displaces exactly its weight in water as it floats. As it melts its volume is reduced to about 9/10 of its original volume, and this smaller volume does not overflow the glass.

When the ice formed it expanded to occupy almost 9/10 of the volume it had in the liquid state, but it did not increase in weight.

If the water is allowed to get warm after the ice melts it should overflow, because it will expand as it warms to room temperature.

Mysterious Ice Cubes

MYSTERIOUS ICE CUBES

NEEDED: Two salad bowls or small baking dishes, a pitcher of ice cubes in water.

EXPERIMENT 1: Fill a bowl almost full of water from the pitcher of ice water. Fill another bowl equally full of warm water. Try to float an ice cube in the center of each bowl.

EXPERIMENT 2: Float two ice cubes in the bowl. Do they move apart or come together?

REASON: The ice cube may be floated in the center of the ice water without difficulty after the water has stopped moving. But it should be impossible for the cube to remain in the center of the warm water.

This is because the warm water melts the ice, the cooler water from the melting ice goes downward, causing irregular convection currents that rise and move to the edges of the bowl. The moving convection currents carry the ice with them to the edge.

Water in the Flame

WATER IN THE FLAME

NEEDED: A flame as from a candle or gas.

EXPERIMENT 1: Hold a container of cold water above the flame briefly, and water will be seen condensing on the cold container surface.

EXPERIMENT 2: Note the white "smoke" coming from an automobile exhaust on a cold day. The white particles are water droplets. Burning of the fuel in the engine produces water, carbon dioxide, carbon monoxide, and other gases.

ADHESION, COHESION

NEEDED: Two squares of wood suspended from a yardstick as shown, water, a pan for the water, sand.

EXPERIMENT: Let the square of wood which is suspended over the water touch it. Pick it up from the water, then balance the squares by putting sand on the other lighter one.

Adhesion, Cohesion

Now let the wet square come down to the water surface. It will cling to the water surface, and a surprising amount of sand will have to be put on the other square to pull it loose.

REASON: The attraction between different kinds of molecules (here, water and wood) is called adhesion. The attraction between molecules of the same kind (here, water) is called cohesion. When the square is pulled away from the water it will be wet on the bottom. It pulled some water up with it. This shows that the adhesion between wood and water is greater than the cohesion between the molecules of water.

THE REYNOLDS RIDGE

NEEDED: A piece of glass, running water, soap.

EXPERIMENT: Clean the glass thoroughly, then hold it under the faucet until the soap has been washed off. Take it away from the water, then before the film of water has drained off, touch the bottom of the glass with a soapy finger. A ridge will be seen to rise up the glass in the water film.

REASON: The clean water has greater surface tension than the water contaminated by the finger at the bottom of the glass. The greater surface tension draws the contaminated water up, making a ridge where the two kinds of water meet. This is called the

The Reynolds Ridge

Reynolds Ridge. It may be seen many places in nature, particularly at the edges of bodies of water.

Oil on the finger tip will cause the ridge to move up, but more slowly. The finger alone usually will contaminate the water enough to cause a slowly-moving ridge. To make a standing ridge in a natural body of water both water and contaminant must be flowing. The ridge moves upstream exactly as fast as the water moves downstream. This can often be seen at a stream bank.

THE HYDRAULIC JUMP

NEEDED: A cookie sheet and the kitchen sink.

EXPERIMENT 1: Let the water pour onto the sheet, and the movement of the water will form one type of hydraulic jump, as shown in the illustration.

EXPERIMENT 2: Watch the water spilling over a dam. It flows rather smoothly to the "jump," then froths, foams, and bubbles in turbulence at the point where its speed decreases.

REASON: The hydraulic jump is a stationary wave very complicated and difficult to analyze. Higher mathematics is required.

The hydraulic jump is a sort of shock wave, similar to the sonic boom emitted by an object traveling faster than sound.

The jump is actually a wave traveling upstream at the charac-

The Hydraulic Jump

teristic velocity of a wave on the water. It is a standing wave because the wave velocity at the point and the water velocity are the same. The wave is traveling upstream at its characteristic velocity relative to the moving water, but it is standing still relative to the land.

SIPHON WITHOUT SUCKING

NEEDED: Two containers, water, a small hose.

EXPERIMENT: Place water in one container; set it on a table. Place the other container on a chair or floor so it is lower than the first.

Hold up the hose, and let it coil into the water so that it fills with water. Place a finger over the end and lower that end into the other container. Remove the finger and the siphon starts.

REASON: Placing the finger over the end of the hose allows the hose to remain full of water as it is lifted over the edge of the container. When the end of the hose is placed lower than the water surface the pressure of the atmosphere on that surface forces water to flow down the hose.

Why doesn't atmospheric pressure act equally on the lower end of the hose? Because the pressure in the lower end of the hose is at atmospheric pressure plus pressure of the column of water in the hose that acts against it.

Don't suck on a hose to start a siphon. Some liquids can be dangerous if sucked into the mouth.

Siphon Without Sucking

Sheet Erosion

SHEET EROSION

NEEDED: A pan of topsoil, some coins or flat stones, rain or a soft sprinkling from a hose.

EXPERIMENT: Place the coins or stones on the soil in the pan, place the pan out in the rain, and watch.

Soil is slowly washed away, and where the coins or stones give protection the erosion is not as noticeable. This type of erosion, hardly noticed, gradually carries away the topsoil from a field. It is called sheet erosion. The best remedy is to keep something growing on the soil at all times. Grass protects the soil from the pounding of the raindrops, and the roots bind the soil together, so that there is little or no erosion.

WATER SPURT

NEEDED: A plastic vinegar jug with cap, water.

EXPERIMENT: Make a small hole near the bottom of the jug. Fill the jug with water, screw the cap on, and the water does not spurt out. Loosen the cap and the water spurts out again.

REASON: This is a matter of surface tension. The weight of the water normally would make the water spurt out, but when no air gets in at the top of the jug the water does not flow because the pressure of the atmosphere against the water at the hole—and the

Water Spurt

surface tension—keeps the water in.

If the hole is large, air can get in through it at the same time water spurts out, and the water will gurgle out. But if the hole is small, surface tension of the water at the hole is strong enough to prevent the flow of water out and air in at the same time.

Capillary Attraction

CAPILLARY ATTRACTION

NEEDED: Eyedropper with a long thin snout, water, a glass.

EXPERIMENT: Note that the water in a glass curves upward where it touches the glass. Now touch the end of the dropper to the water surface. The water not only rises where it touches, but moves slightly upward into the dropper.

REASON: The same thing happens in both cases, but the small end of the dropper allows the surface tension to pull up a little water not only at the edge of the container but throughout the container's width.

A small glass tube is called a capillary tube, from a Greek word meaning "hair." The smaller the tube the higher the capillary action will lift the water. A cloth or piece of rough paper touched to the water surface shows capillary action; it is made up of small fibers that make the water rise between them.

Clinging Water

CLINGING WATER

NEEDED: A saucepan, a bowl of larger diameter, a water hose.

EXPERIMENT: Pour water into the bowl. When full, the water will flow down the sides of the bowl until the pan is full, then will flow down the sides of the pan. The water clings to the bowl and pan, some of it flowing out on the handle of the pan before flowing to the sink or ground.

REASON: Adhesion, here, is attraction between unlike molecules, those of the water, and those of the objects. The attraction causes the water to cling to the pan and the bowl. Cohesion causes the water molecules to cling together.

There are many interesting and simple surface tension adhesion and cohesion experiments. Molecules of water in a container cling together so that the water surface will support small steel objects such as needles and razor blades. This is an example of surface tension experiments.

THE DRIP

NEEDED: A water faucet (spigot or tap mean the same) or an eyedropper. Water.

EXPERIMENT 1: Open the spigot a little; water will pour

The Drip

downward in a narrow stream. Close the spigot slowly; a point will be reached where the water will not flow, but will drip.

EXPERIMENT 2: The same experiment may be performed with an eyedropper.

When the flow is decreased in the right amount, the surface tension squeezes the water into droplets as they leave the spigot surface. The drawing may help to show this.

Suds

SUDS

NEEDED: Two jars, detergent, the kitchen sink.

EXPERIMENT 1: Put a half teaspoonful of detergent in both jars. Let water run slowly into one; let it run fast into the other. The second jar will fill rapidly with suds that overflow.

EXPERIMENT 2: Try this with hot water.

REASON: The rapidly moving water breaks through the surface of the water in the jar, carrying air with it. This air "blows bubbles" which increase the volume of liquid in the jar and also decrease its density. As more water comes in and more bubbles are formed they float on the surface of the water, rise, and some of them overflow.

The slow-moving water in the other jar carries very little air with it into the detergent solution, and so does not blow as many bubbles.

Water Cavities

WATER CAVITIES

NEEDED: Water in a tub or pool, a large spoon.

EXPERIMENT: Move the hand or spoon quickly through the water. If it is moved fast enough, a bubble resembling air will be seen behind the hand or spoon.

REASON: Pressure behind the moving spoon or hand is reduced—a partial vacuum is produced, causing the water to go through the same phenomenon as boiling. Water can boil from heat, also it can boil from reduced pressure. Boiling occurs when the vapor pressure of the water is equal to the surrounding pressure.

This effect, called "cavitation," is damaging to propellers on large boats, and to large metal tubes through which water flows, such as spillway tunnels at dams. In clear water, cavitation can be seen clearly in the wake of boat propellers.

In the drawing the spoon is being moved in the direction the arrow points.

Liquid Layers

LIQUID LAYERS

NEEDED: A tall glass, water, alcohol (rubbing alcohol will do), cooking oil, syrups of different thicknesses.

EXPERIMENT: Fill the glass half full of water. Add about a third that much alcohol—the water and alcohol will mix. Add a little of the thickest syrup, then a little of the thinner syrup. They will sink to the bottom of the glass and remain in layers. Finally, add cooking oil; it will float on top.

REASON: Specific gravity is the weight of a substance compared with the weight of an equal volume of water. Liquids with specific gravities higher than that of water, which is 1, will sink in water; liquids with specific gravities lower than that of water will float on water unless they mix with it, as did the alcohol.

Alcohol has a lower specific gravity than water, so when it mixes with water the specific gravity of the mixture is lower than 1. From the bottom upward the lighter liquid layer floats on the next heavier layer below it if they do not mix or dissolve in one another.

Sand Castles

SAND CASTLES

NEEDED: A sandy beach.

EXPERIMENT: Build a sand castle, and note how the sand particles hold together when wet yet fall apart when dry.

REASON: Water molecules pull to each other in all directions when there is water in all directions. When there is water only on three sides, however, the pull is much like a stretched rubber sheet. We call this surface tension. It is this surface tension that holds the wet sand particles together.

If the sand is dry there can be no surface tension. Sand under water does not have a surface tension effect because the attraction between molecules is equal in all directions. The sand castle can be built when all grains of the sand are attached to all other grains by the surface tension of the water between them.

Fluids are usually considered to be either gases or liquids. Yet other substances act like fluids, including dry sand, flour, or even gravel as it is poured from a truck.

(Suggested by Pierre LaFrance, in the magazine *The Physics Teacher*, January 1975).

Dry Solids—Contact Angles

DRY SOLIDS—CONTACT ANGLES

NEEDED: A jar of wet sand, a jar of dry sand.

EXPERIMENT: Pour a little water into each jar. The water will sink quickly through the wet sand, more slowly through the dry sand.

REASON: A drop of water on a solid surface forms a "contact angle" which can be zero, in which case the water does not wet the solid at all. If the contact angle is 180 degrees the liquid wets the solid perfectly. If the solid surface has been wet, then we must think of the contact angle of the fluid with itself, which is 180 degrees.

In the dry sand the water forms little arches across the grains at the characteristic contact angle of water with silicon dioxide or sand. When the sand has been previously wetted the water runs through freely, following the films already established by the wetting.

Note: There is a resistance to wetting which requires a small amount of energy to overcome.

Rising Water

RISING WATER

NEEDED: A glass of water, a piece of paper towel.

EXPERIMENT 1: Sight along the surface of the water in its glass. It is a little higher where it meets the glass than elsewhere. Adhesion is the force that holds the water to the glass, causing the water to rise against the surface of the glass.

EXPERIMENT 2: Dip the end of a strip of paper towel into the water. The water will rise through the fibers of the paper, higher than it did at the edge of the glass, but for the same reason. The rise of water in the narrow spaces between the paper fibers is called capillarity, from the Latin word capillus, meaning hair.

REASON: Movement of the fluid in capillarity is not always up. It may be down. If a narrow glass tube is pushed into mercury the part of the mercury next to the glass is depressed. Glass attracts water more than water attracts water; mercury attracts itself more than it is attracted by the glass.

Clear Ice Cubes

CLEAR ICE CUBES

NEEDED: Water, ice trays.

EXPERIMENT 1: Fill one ice tray with water from the spigot. Fill the other with water that is almost boiling hot. Place both trays in the freezer. Ice cubes made with hot water should be clear.

EXPERIMENT 2: Put the tap water into a jar and shake it vigorously, then put it into an ice tray. Shaking should put more air into the water.

EXPERIMENT 3: Let a jar of hot water cool, then shake it. The water then should make cubes full of bubbles. Shaking dissolves air into the water.

REASON: All water has some air dissolved in it. Water direct from the spigot has a considerable amount; boiling or heating releases most of the dissolved air, which comes off as bubbles. Freezing also releases dissolved air, but instead of bubbling off, it stays in the cooling water, allowing ice to freeze aound it. It is this trapped air in ice cubes that accounts for most of the cloudy or streaked appearance of the cubes.

A WATER UMBRELLA

NEEDED: A water hose, a jar with a smooth lid.

EXPERIMENT 1: Direct the water stream onto the jar lid.

A Water Umbrella

With practice the water can be made to form an "umbrella" as shown.

EXPERIMENT 2: Hold a spoon under the flow from the kitchen sink faucet. The umbrella can be formed.

REASON: Surface tension keeps the water in a sheet as it flows over the edge of the jar top. As the sheet gets thinner and falls at a greater speed the surface tension pulls it into drops.

Surface tension acts as a rubber sheet on the surface of water, tending to draw it into the shape having the smaller surface area.

SWELLING FRUIT

NEEDED: Dried fruit, water, salt, two bowls.

EXPERIMENT 1: Put salt water in one bowl and plain water in the other. Place dried fruit: apricots, prunes, apples for example, in both bowls. In 12 hours the fruit in the plain water will swell. Fruit in the salt water will not.

EXPERIMENT 2: Salt is a preservative. Put small cucumbers in strong salt solutions; the living cells that may lead to decay are dehydrated as the salt solution draws their liquids out through osmosis. The cells cannot live.

REASON: In osmosis water flows both ways through cell membranes. But the flow is greater toward the more concentrated solution, which in this case is inside the dried fruit. The flow of

Swelling Fruits

water into the fruit causes it to swell.

If salt water is tried there is likely to be no flow, since the concentration inside and outside of the cells of fruit should be about the same.

CLOUD IN A JUG

NEEDED: A milk jug and a match

EXPERIMENT: Hold the lighted match under the mouth of the jug for a few seconds, then blow hard into the jug, compressing the air in it as much as possible. Release the air pressure suddenly, and a cloud will appear in the jug and remain for an instant.

REASON: Blowing into the jug compresses the air and adds a little moisture from the breath. The compression heats the air slightly, but most of the heat is absorbed by the glass walls of the jug. As the pressure is released, the expansion of the air cools it, and since it cannot then hold as much moisture as the warmer air, some of the moisture is condensed briefly into droplets that make up the cloud.

Cloud In a Jug

The lighted match warms the air in the jug slightly, and adds tiny particles of smoke around which water vapor molecules can condense to form the fog. Putting particles into air for this purpose is called "seeding" the air. The invisible separate molecules of water vapor condense when cooled on the solid particles to make the visible fog.

Repeat the act of compressing and releasing the pressure in the jug. The cloud becomes more dense each time if this is done quickly.

THE SPIRALING WATER

EXPERIMENT: Notice the direction of the spiraling of the water as it goes down the drain. Contrary to the statement sometimes heard that it always whirls clockwise in the southern hemisphere and counterclockwise in the northern hemisphere, it can be seen to whirl either way.

The turning of the earth on its axis can start the spin in agreement with the statement, provided there is no motion in the water to begin with. But a very slight motion in the water can determine which way it will whirl. The vortex seen is due to gyroscopic action in the water. Look up "Coriolis Effect."

PHYSICAL EQUILIBRIUM

NEEDED: A large jar, a little hot water, a cover, perferably glass, for the jar.

Spiraling Water

EXPERIMENT: Cover the jar with the hot water inside, and watch what happens. The water tends to evaporate, yet the vapor condenses on the sides of the jar and the top, changing back into water. Physical equilibrium is defined as that situation in which the liquid molecules evaporating and vapor molecules condensing are equal in number.

"MAGNETIC" SUGAR

NEEDED: A bowl of water, a wooden match or toothpick, a sugar cube.

EXPERIMENT: Float the wood on the water. Dip a corner of the sugar cube into the water near the end of the wood, hold it there, and the wood will begin to move toward the sugar. Move the sugar away slowly, still keeping it in the water, and the wood will follow it as if drawn by magnetic force.

REASON: There is no magnetism involved, of course. The dissolving sugar makes the water heavier at that point, the heavier water flows downward, causing a current in the water. The wood is moved by the water that flows in to take the place of that which has gone down.

If the cube is touched lightly to the water just in front of the wood, the wood is likely to move away. This is because the sugar beginning to dissolve in the water reduces the surface tension at that end of the wood. The surface tension at the other end is as strong as before, and like a stretched rubber sheet, pulls the wood back. But as more sugar dissolves and begins to fall through the water, the flow begins that will move the wood toward the sugar cube.

SMOKE RING IN WATER

NEEDED: A dish or jar of water, a drop of ink or colored water.

EXPERIMENT: Drop the ink into the water, and in many cases it will form into rings resembling smoke rings.

REASON: As in the formation of vortex rings with the Smoke Cannon, there is a drag where the outer edges of the ink drop move through the water, causing the ink to form a toroidal or doughnut shape. Clear water pulled into the middle by the turning effect keeps the center of the doughnut more or less clear.

The Crooked Water Surface

THE CROOKED WATER SURFACE

NEEDED: A glass of water and a small lid that will float.

EXPERIMENT: Fill the glass almost full. Float the lid on the surface of the water, and it will move to the edge of the glass. Fill the glass completely, and the lid will float in the center of the water surface.

REASON: Notice that when the glass is not filled, the water surface between the lid and the glass is bent upward as in diagram A. The surface tension of the water, acting like a stretched rubber sheet, draws the lid to the nearest part of the glass.

When the glass is filled, the water surface is in the shape shown in diagram B. Here the surface tension acts to push the lid away from the glass, with a similar effect at all points on the glass.

WET MUD

NEEDED: Observation only. We know that mud will not brush off of our clothes, but will brush off after it dries.

REASON: The surface tension of water binds the clay particles into mud. Without this bond, the dry loose particles easily brush apart. The clay particles are also bound to the cloth by surface tension.

Surface Tension

SURFACE TENSION

NEEDED: Two dishes of water, two Ping-Pong balls, a small piece of soap.

EXPERIMENT: Place a ball on the surface of the water in a dish, and it will move to the nearer side and cling to the dish. Mix some soap in the water in the other dish, place the ball on the water surface, and it will remain in the center of the dish.

REASON: The water molecules attract the ball and the dish. The effect, surface tension, can be seen as the curved shape of the water surface. The double effect of the surface tension causes a small force to pull the ball to the nearest side of the dish. Soap in the water greatly weakens the surface tension, so that the curving effects are not noticed. The water surface will appear flat, and there will be insufficient force to pull the ball to a side.

THE "LIVE" WIRE

NEEDED: A piece of fine wire such as that found in lamp cords, a bowl of water, a drop of oil.

EXPERIMENT: Wind the wire into a flat spiral shape, place it on the water surface carefully so it will float, and let the water become still. Place a drop of oil in the center of the spiral, and the wire will "come alive" and begin to turn.

REASON: The oil floats on the surface of the water and tends to spread out evenly over that surface, reducing the surface tension beginning at the center and spreading outward. There is a slight force on the wire, and this makes the spiral turn.

The Rising Bubble

THE RISING BUBBLE

NEEDED: Soap solution and a glass or plastic funnel.

106

EXPERIMENT: Dip the funnel into the soap solution and withdraw it, so that a film of soapy water is lifted with it. The film will be at the large end of the funnel, but will crawl slowly up to the little end as the funnel is held as shown. (The funnel must be clean and wet with soapy water for this to work successfully.)

REASON: The film has two surfaces, both of which act like stretched rubber sheets (surface tension is the term for this). They tend to shrink to the smallest surface area, and their force due to surface tension is great enough to lift the film upward against the pull of gravity. Dirt tends to reduce the adhesive forces due to surface tension.

A Surface Tension Puzzle

SURFACE TENSION

NEEDED: Two toothpicks, a bowl of water, a small piece of soap.

EXPERIMENT: Place the toothpicks side by side on the still water surface, and they tend to move together. Touch the soap to the water between them, and the toothpicks move apart.

REASON: Soap and many other substances will reduce the surface tension. If the soap is placed between the toothpicks the surface tension there is reduced, allowing the greater surface tension effect on the outside of the toothpicks to pull them farther apart.

Dip a toothpick into alcohol, and touch it to the water surface between the floating toothpicks. The floating toothpicks will spring apart for the same reason. Alcohol, too, reduces the surface tension.

A SURFACE TENSION PUZZLE

NEEDED: A bowl of water, an ice cube, two matches or toothpicks.

EXPERIMENT: Let the water become still, then place the matches side by side as shown. Touch the ice cube to the water between the matches. The cooling of the water increases the surface tension and should bring the pieces of wood together. But it does not work this way. The matches move apart.

REASON: Water currents set up by the melting ice float the matches apart. The effect of these currents is stronger than the pull due to the increased surface tension of the slightly cooled water. The water currents move both downward and laterally.

HOLES THAT HOLD

NEEDED: Small mesh kitchen strainer, water, oil.

Holes That Hold

EXPERIMENT: Pour water down the side of the strainer, and it runs through. Dip the strainer into oil, sling out the excess, then try the water. The water will collect in the bottom, and not run through the mesh until a considerable amount has collected in the bottom of the strainer.

REASON: The surface tension or molecular attraction is strong enough between the meshes of the wire to hold water. If the wire is not oily the water can "wet" it and flow easily on its surface to the under side, then spill from the strainer.

While the strainer is holding water, touch a finger to the underside of the mesh. This breaks the surface tension and allows the water to flow through to empty the strainer.

THE MOVING TOOTHPICK

NEEDED: A toothpick, some soap and some water.

EXPERIMENT: Stick one end of the toothpick into the soap, pull it out, then float the toothpick carefully on the water surface. It moves about mysteriously.

REASON: The soap decreases the surface tension of the water at that end of the toothpick, and the stronger pull due to greater surface tension at the other end causes the toothpick to move along. The water surface must be calm.

Buoyancy

BUOYANCY

NEEDED: A tin can with both ends cut out, a cardboard square, a bucket of water.

EXPERIMENT: Completely cover one end of the can with the card, and push the can and card down into the water. The buoyancy of the water will keep the card against the end of the can. Then slowly pour water into the can. The card stays on until the water inside the can reaches the height of the water outside it, in the bucket, or until the can and card are completely submerged. Then the card falls off.

REASON: Pressure beneath the surface of water at any particular depth is equal in all directions, and when there is no water in the can the water pushes against the bottom of the card to hold it in place. As water is poured into the can the pressure on the bottom of the card decreases, and when the water inside reaches the level of water outside the can there is no pressure on the bottom of the card in excess of that above. The card falls off in response to the pull of gravity.

Chapter 7

Gravity & Centrifugal Force

WEIGHTY PROBLEMS

NEEDED: A scale, a heavy weight.

EXPERIMENT 1: Stand on the scale and raise one foot. Do you weigh less?

EXPERIMENT 2: Lift the weight with one hand, then with two. It seems much lighter when lifted with two hands.

REASON: (1) When you stand on two feet, your weight is applied to the scale platform through both. Lift one foot, and its weight is transferred to the other, and through it to the scale. The reading is the same.

(2) When the weight is lifted with two hands, each hand has to support only half as much as if it is lifted with one hand. The weight is the same, but feels less when two hands divide it. This is simply a psychological illusion.

A BIRD MYSTERY

NEEDED: A wooden box with a perch, a bird, a scale.

EXPERIMENT 1: Watch the weight scale. There will be some variation as the bird leaves the perch and returns to it, but the *average* weight of the bird on the perch and in the air inside the box will be the same.

EXPERIMENT 2: In an open cage, the bird's weight in flight may register less because part of the downward movement of air caused by the flying will go outside the cage and will not register on the solid bottom.

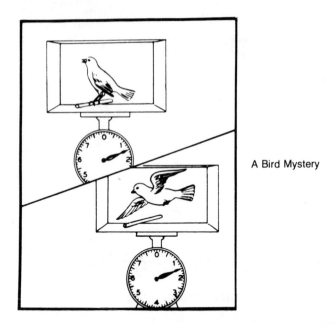

A Bird Mystery

REASON: When the bird is flying, its weight is held by the air, and the downward push is transmitted through the air to the bottom of the box. Here, too, there will be some variation in the weight recorded, but the average will be the same as when the bird is at rest.

Chapter 8

Electricity & Magnetism

EARTH MAGNETISM

NEEDED: A wire coat hanger, string, a paper clip, a compass.

EXPERIMENT 1: Straighten the coat hanger, and hang it balanced on a string or small thread. This must be done in a room where there are no people and no air movements. Overnight the wire will be found pointing approximately north and south like a compass.

REASON: The wire, while not magnetized, still tends to line

Earth Magnetism

up with the earth's magnetic lines of force. It is soft iron, and will not remain permanently magnetized, as steel would.

EXPERIMENT 2: Move slowly, so as not to produce air currents. Hold a paper clip or other iron object near the end of the wire, and the wire will be attracted to it.

EXPERIMENT 3: Use the compass to test large iron or steel objects in the house, such as a bathtub or refrigerator. They will be found to be magnetized simply by setting in one position in the house. Their magnetism comes from the earth's magnetic lines of force.

An Easy Battery Tester

AN EASY BATTERY TESTER

NEEDED: An "alligator" clip, piece of wire, a flashlight bulb.

EXPERIMENT: Attach a wire to the clip, place the bulb in the clip, and you have a battery tester. Hold the bare end of the wire against the bottom of the battery, touch the tip of the bulb to the metal at the tip of the battery, and if there is current, the bulb will light.

COMMENT: Do not try this on batteries of higher voltage than the bulb rating. This is suggested for simple flashlight cells, using a regular 1 1/2-volt or 1.3-volt bulb. To test nine-volt batteries a nine-volt bulb must be used, but it is easy to change bulbs.

There is no danger; touch the metal parts of batteries freely. There is no shock. Do not use this on house current.

There are batteries that produce voltages as high as 90 or even 300. There can be shock from them; a 300-volt battery can be dangerous. These are not found in homes or schools, however.

A Homemade Compass

A HOMEMADE COMPASS

NEEDED: A sewing needle, a magnet, felt, styrofoam, a dish of water.

EXPERIMENT 1: Rub one end of the needle against one end of the magnet a few times.

This will magnetize the needle.

Break off a piece of styrofoam the size of a pea, stick the needle through it, and float it on the surface of the water. It will move until the needle points north and south.

EXPERIMENT 2: Cut a piece of felt half an inch wide and as long as the needle. Place the needle on it, and it will float on the water for days if the water is not disturbed. This is another way to make the compass on water.

REASON: A magnetized piece of steel will tend to align itself with the lines of force from any magnet. In this case the magnetic lines of force are those of the earth, which run approximately north and south, and line up the needle with them.

A needle without styrofoam or felt may float on the surface of water, held up somewhat precariously by the surface tension of the water. The felt and styrofoam do not wet easily, and as long as they are not wet they float on the water.

The Electric Girl

THE ELECTRIC GIRL

NEEDED: A girl with long hair, a board, pint Mason jars, orlon acrylic garment, a neon lamp (Calectro NE-51 will do), two more people.

EXPERIMENT: Make a platform with the board and glass jars. Have the long-haired girl stand on it. Have someone rub her hair downward to the waist with the acrylic. The third person may then draw a spark from the insulated girl's finger.

REASON: Rubbing the hair puts a charge of electricity over the girl on the platform. The charge (or part of it) will jump as a spark to someone else. This works only when humidity is very low, usually only in cold winter in a heated-room. But it is possible to get 10,000 harmless volts this way.

If the girl holds a neon lamp, and someone else takes the spark from the tip of the lamp, it will flash over and over, each time the spark jumps.

The jars must be clean and dry. The hair and the acrylic must be dry, and the rubbing must be vigorously done. Note how the hair tends to stand out; each hair takes on the same charge, and like charges repel.

It may be more convenient to hold the lamp bulb with a wire handle as shown, but this is not necessary. The high voltage involved is perfectly harmless; the electric current is very low.

What Conducts Magnetism?

WHAT CONDUCTS MAGNETISM?

NEEDED: Magnet wire wound around a cardboard tube, battery, compass, objects to test.

EXPERIMENT: Place the tube near the compass, at right angles to the needle. Touch the wires to the battery, and the needle is deflected. Move the tube away until there is only a slight deflection.

Now try inserting various objects into the tube, and see whether the deflection is stronger. Try copper wires or rods, wood, nails, aluminum. The nails, which are iron, will be found to conduct the magnetism, and the needle deflection should be quite strong.

REASON: There is an electromagnetic field in and around the tube when electric current flows in the wire. Materials other than iron are not affected by the field, nor do they affect it. Iron materials themselves become magnetized in the field and make the total magnetic field stronger.

A hard steel object, such as a file, may remain magnetized after the current is turned off.

WIGGLE WIRE

NEEDED: A magnet (permanent type), toy train transformer, wire, screws, a wood base.

EXPERIMENT: Mount the magnet on the base, and stretch a wire (not tightly) between screws, a half inch from the pole of the magnet as shown.

Wiggle Wire

Connect one end of the wire to the transformer. As the other end of the wire is touched to the transformer the wire will be seen to vibrate.

REASON: Current carried by the wire sets up a magnetic field around it. It is attracted or repelled by the fields from the permanent magnet according to the rule "like poles repel and unlike poles attract."

Since the alternating current from the transformer (don't use dc) reverses its direction 120 times a second, the wire is alternately attracted and repelled as the magnetic field around the wire is rapidly reversed.

Either a U-shaped magnet or a bar magnet may be used. The wire used by the author in his model was 20 gauge copper magnet wire.

CRAZY PENDULUM

NEEDED: Two bar magnets and string.

EXPERIMENT: Suspend a magnet on a string as shown, and hold it over the other magnet. If they attract, reverse the lower magnet. Then bring them close, and the motion of the upper magnet will be erratic and "crazy."

REASON: The similar magnetic fields repel, and the repul-

Crazy Pendulum

sion is greater as the magnets are close. They move apart, and the repulsion is less, so they come close again.

The string allows the upper magnet to move in many directions.

There is a device marketed commercially which operates on this principle. One magnet is in the base, and the other in a sphere suspended above it. It is sold as an interesting conversation piece.

The animals shown in the illustration have bar magnets attached to their feet. They are dime store magnets.

A PENCIL MAGNET

NEEDED: Insulated copper wire, a large nail, a pencil, paper clips, dry cells or a toy train transformer.

A Pencil Magnet

EXPERIMENT 1: Wind 50 turns of wire around the nail, hold it over the pile of paper clips, connect the ends of the wire to the power source, and a magnet is formed that will pick up the clips, if they are made of iron.

EXPERIMENT 2: Wind 50 turns of the wire around the pencil and try to pick up the clips. Again we have an electromagnet, but a much weaker one.

COMMENT: A popular school science book states that a magnet cannot be made using the pencil. This experiment shows that it can. The difference is that the iron nail concentrates the lines of force from the wire and focuses them to the end of the winding.

A Loop Magnet

A LOOP MAGNET

NEEDED: Iron filings or bits of steel wool, small gauge copper wire, a six-volt lantern battery.

EXPERIMENT: Bend the wire as shown, and connect one end to the battery. Touch it to the filings, and nothing happens. Hold the other end of the wire to the battery, touch the wire to the filings, and some of them are picked up as the wire is lifted.

REASON: There is a magnetic field around any wire carrying a current. Since iron and steel are attracted by a magnet, they are lifted, because we have made a magnet.

In many books of experiments it is suggested that only a straight wire can be used for this. It may be a difficult experiment with only a wire. But if one turn is put into the wire, as shown, it is easy.

Do not hold the wire against the battery terminal more than a few seconds. It would run the battery down rather quickly if held for very long.

The drawing shows a toy transformer used instead of a battery. It is better because it is stronger. But the wire may get hot, so wear a glove on the hand holding the wire.

Direction of Magnetic Fields

DIRECTION OF MAGNETIC FIELDS

NEEDED: Battery, wire, a compass.

EXPERIMENT 1: Wind some of the wire into a spring shape. Place the compass close (or better still, under the straight wire) then touch the ends of the wire to the battery. The needle will be deflected.

Place the compass beside or under the spring part of the wire, touch the ends to the battery again, and the needle will be deflected in the opposite direction.

EXPERIMENT 2: Try this with the compass under the wire, then with the compass over the wire.

REASON: Lines of force are circular around the wire, with the wire at the center of the circle. The compass needle lines itself up with the earth's lines of force, which are north and south, but the lines from the wire are stronger and can cause the needle to turn at right angles to the wire.

One cell of flashlight battery is sufficient for this experiment.

Bubble Magic

BUBBLE MAGIC

NEEDED: Bubble blowing device, a rubber balloon, wool or fur.

EXPERIMENT 1: Charge the inflated balloon by rubbing it with wool or fur, then have someone make a bubble, release it from the plastic bubble maker, then let it fall again to the plastic. Move it close to the charged balloon, and it will be distorted as shown.

EXPERIMENT 2: Release the bubble into the air, bring the charged balloon near it, and the balloon will attract the bubble, causing it to move toward the balloon.

REASON: The negatively charged balloon causes some negative charges on the bubble to move away to the side farthest from the balloon. This leaves the side of the bubble nearest the balloon with a positive charge. Thus the two attract each other. The thin soap film is easily distorted by the slight pull of the static charge on the balloon.

(Remember that static electricity experiments may not work unless the humidity in the room is low.)

A SENSITIVE STATIC DETECTOR

NEEDED: Light cardboard, a needle, scissors, a small bottle, a comb.

A Sensitive Static Detector

EXPERIMENT: Cut out the cardboard as shown, and stick a needle through the middle so the point is just below the edge of the card. Rest the needle point on the bottle top. If it does not balance, snip off small bits of card on the heavy side until it does.

Rub the comb on wool or almost any other kind of cloth to charge it. Hold it near the end of the balanced card, and the attraction or repulsion will make the card revolve.

REASON: The rule is: like charges of static electricity repel each other; unlike charges attract. Any charged article will attract any uncharged article such as the cardboard. Since there is practically no friction involved in the turning of this device, very small charges will turn it, even a hand rubbed on clothing on a dry day.

BOARD MAGIC

NEEDED: A light-weight six-foot or eight-foot board, 1 × 3 or 2 × 4 will do, a comb, something to rub the comb with to produce a charge. Glass rubbed with silk or a rubber or plastic comb rubbed with fur or through the hair will do.

EXPERIMENT: Balance the board on a string. Dr. John A. Davis, of Kansas State University, balances the board on a watch glass. Bring the charged glass rod or comb near the end of the board without allowing it to touch. Attraction between the comb and board will cause the board to turn slowly.

REASON: Like charges repel each other; unlike charges attract. The charged comb and neutral board attract because when they are brought close to each other the charged body causes like

Board Magic

charges in the originally uncharged body to leave that area, thus leaving the board with an opposite charge. Now the two bodies attract each other.

Static experiments are usually done with small objects. Use of a large pole makes this variation rather amazing. It will work only where humidity is low, and where there is no draft in the air.

MAGNETIC FIELDS

NEEDED: Iron filings, a magnet, and a piece of glass.

EXPERIMENT: Place the piece of glass over the magnet. Drop the iron filings onto the glass and watch what happens. The filings align themselves into a pattern showing the magnetic lines of force (magnetic field) of the magnet.

COMMENT: Iron filings may be obtained from a garage mechanic. Use the tiny iron pieces that fall to the floor when brake drums are turned. Clean them with hot water and detergent and dry them in an oven. To prevent rust, give the filings a last wash in rubbing alcohol to which a few drops of light machine oil has been added.

Steel filaments may be used instead of the filings, which can only be used once or twice. These can be obtained by rubbing two steel wool wads together, or by rubbing steel wool between sandpaper blocks, as shown in the drawing.

When handling steel wool, be careful not to breathe the particles or get them stuck in the skin. Never tear a piece of steel wool with the bare fingers; use scissors to cut off what you need.

Magnetic Fields

GLOWING IN THE DARK

NEEDED: A fluorescent lamp in a darkened room.

EXPERIMENT: 1 Have the light on. Cover the eyes with a hand to shut out the light. Count off 15 to 25 seconds, switch off the light, and open the eyes quickly. The tube will be seen glowing, the glow gradually but quickly fading away.

REASON: The light given by a fluorescent tube is produced

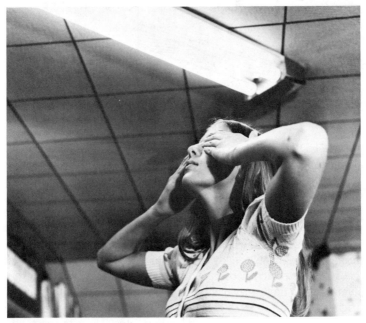

Glowing in the Dark

when the electric current flows through a gas containing a small amount of mercury. The light produced is ultraviolet, not suitable as room illumination.

But the ultraviolet light shines on a coating inside the tube, a coating made of phosphors, substances that glow with different colors when excited by the ultraviolet light. The ultraviolet goes out as the switch is turned off, but the phosphors continue to glow briefly.

Different phosphors give off different colors, and in the tube, phosphors are used that emit light visible to the eye and suitable for illumination.

EXPERIMENT 2: Try the same experiment with a television set. Its picture is made by phosphors that can be seen to glow briefly after the set is turned off.

Aluminum Magic

ALUMINUM MAGIC

NEEDED: A strong magnet, round pencils, a straight strip of aluminum.

EXPERIMENT: Place the aluminum on the pencils so it can roll easily on them. Hold the magnet just above the strip, and move it quickly back and forth. Do not let it touch the aluminum. Yet the aluminum will move.

REASON: As the magnetic lines of force from the magnet cut through the aluminum, eddy currents of electricity are produced in the strip. An electric current produces magnetic fields. In the aluminum the fields produced interact with the fields of the strong magnet, and pull or push the strip back and forth.

This is the principle of the induction motor. There the currents

Permanent Pictures of
Magnetic Fields

induced in coils are strong enough to turn the armature of the motor and do work.

An improved way to do this is shown in the drawing. Place a piece of window glass on strips of wood so it rests just above the aluminum. Then rub the magnet back and forth on the glass. This way the magnet does not touch the aluminum yet stays close to it.

PERMANENT PICTURES OF MAGNETIC FIELDS

NEEDED: A magnet, some white cardboard, iron filings, an atomizer or other water sprayer, salt, a can of aerosol spray paint.

EXPERIMENT: Place the card over the magnet, sprinkle the iron filings on the card, tap lightly and they will assume a pattern representing some of the lines of force of the magnet.

There are two easy ways for preserving the patterns.

1. Spray salt water on the iron filings from the sprayer. Repeat after a few hours and let the experiment stand overnight. Next morning the outline of the filings will remain on the card as rust, and the filings will fall off the paper when the magnet is removed.

2. Set up the card with iron filings as before, but this time spray the card lightly with a dark color paint from the aerosol can.

When the paint dries, we have a permanent record of the pattern of the iron filings. Brush off any loose iron filings.

The paint spray method leaves the filings pattern in white on a dark background, the salt water makes rust on white.

Acrobatic Rice Puffs

ACROBATIC RICE PUFFS

NEEDED: Window glass, a silk cloth, puffed rice, some books, a dry day when the humidity is low.

EXPERIMENT: Place books under the edges of the glass so it rests about 3/4 inch above the table. Spread puffed rice under the glass. Rub the glass briskly with the cloth, and the rice will jump about. Draw a finger along on the glass, and some of the rice puffs will tumble about under the finger.

REASON: The cloth rubs away some of the surface electrons from the glass, leaving it with a positive charge of electricity. Electrons in the puffed rice will be attracted by the positive charge on the glass, and some electrons will move in the rice to the upper surface closer to the glass. The attraction between the positive glass and the negative rice puffs will then be sufficient to make them move about, and jump to the glass. The moving finger will rearrange some of the charge on the glass.

Try putting some of the rice grains in a glass pie dish. Stretch plastic over the dish, and rub it with a finger or a woolen cloth.

AN ELECTRIC PENDULUM

NEEDED: Two tall tin cans, three pieces of paraffin wax, a small thread, a paper clip, a rubber balloon, wool cloth, a dry day when the humidity is low.

EXPERIMENT: Set each can on a paraffin block, suspend the

An Electric Pendulum

clip between them on the thread from the paraffin block as shown below. Rub the balloon with the cloth, bring it close to one of the cans, and the clip will move back and forth between the cans.

REASON: Some of the electrons from the cloth rub off onto the rubber, giving the rubber a negative charge. When it is brought near the can, electrons in the can are repelled to the far side of the can, giving it a charge.

The neutral clip is attracted to it, takes on a similar charge, is repelled from it, and attracted to the other can. There it gives up its charge to the second can, and is attracted again to the first. The ferry-like motion is repeated.

The paraffin serves only as support and insulation, but is very important because of its insulating property.

ELECTRICITY FROM HEAT

NEEDED: Two wires of different metals, such as iron and copper, an alcohol or gas flame, an earphone.

EXPERIMENT 1: Connect the wires to the earphone, and rub their ends together in the flame. Sounds will be heard in the earphone, showing that a small electric current is produced when the wires are hot and touch each other.

REASON: Each different metal has its own natural rate for

Electricity from Heat

losing electrons. Heat increases the loss rate for one metal more than for the other, the more active metal losing some electrons from its surface to the less active metal as contact is made. This transfer of electrons is the electric current. A pyrometer uses this principle to indicate high oven temperature.

EXPERIMENT 2: Dip the ends of the wires into salt water and rub them together while in the solution. Here electricity is produced, not by heat but by chemical action.

MAKE A BIG MAGNET

NEEDED: A compass, a crowbar, a hammer.

EXPERIMENT: Hold the crowbar pointing north and south, with the north end pointing slightly downward. Hit it several times with the hammer. It will become magnetized and will pick up paper clips or small nails.

REASON: The earth acts as a huge magnet. When the bar is held in alignment with the earth's magnetic lines, and struck, it becomes magnetized as clumps of molecules called "domains" line up, with the north end of one near the south end of the other. If the bar is allowed to lie in a north-south direction it will become magnetized after several days. Hitting it allows the domains to align themselves more quickly. The compass will indicate a weaker charge of magnetism on the bar.

STATIC ELECTRICITY AND THE TV

NEEDED: A cotton string, a TV set, a day when the humidity is very low.

EXPERIMENT: Let the string hang about an inch in front of the TV screen. Turn the set on. As the voices and pictures appear the string moves, either attracted to or repelled from the picture tube. Sometimes it is repelled and then attracted. The action stops after a few seconds. It can take place again as the set is turned off.

REASON: The television picture is made by a stream of electrons that hits the phosphor inside the face of the picture tube. When the set is first turned on the electric charge builds up on the inside face of the tube—a static charge that can attract or repel a string. Some of the charge is soon drawn away by the positive charge on the second anode inside the tube. This charge can be due to 15,000–25,000 volts used in the tube. It is of course harmless because it is never released from the tube in dangerous amounts.

THE WILLIAM GILBERT EXPERIMENT

NEEDED: A strong bar magnet, small iron wires that may be

The William Gilbert Experiment

cut from paper clips (if the magnet is quite strong whole clips may be used).

EXPERIMENT: Place the wires carefully along the magnet. They will stand up at various angles in a regular pattern.

REASON: The compass needle "dips" at various parts of the earth, showing the paths of the magnet lines of the earth. The magnetic field of the bar magnet is similar to that of the earth in this respect, and the wires take positions on the magnet representing those of the compass needle on the earth.

Gilbert fashioned a ball of lodestone, a natural magnet, to represent the earth and placed iron wires over its surface. This would make a good science fair project perhaps, but might be difficult. The bar magnet shows it in a simpler way.

STATIC FUN

NEEDED: Old phonograph record, woolen cloth, a ping-pong ball.

EXPERIMENT: Rub the record briskly with the wool. Hold it near the ball, and the ball will be attracted to it.

REASON: Rubbing the record with the wool gives the record an excess of electrons (negative charges). When it is brought near the uncharged ball, the negative charges on the ball move to the far side, leaving positive charges on the side near the record. The attractive force between the record and ball is sufficient to pull the ball along on the table.

Touch the ball with the record, giving it the same charge as the record, and it should be repelled. The rule is: similar charges repeal, unlike charges attract.

MAGNET IN THE BATHROOM

NEEDED: A compass.

EXPERIMENT: Hold the compass at different places against the bathtub. The needle will turn, so that it indicates a south pole at some points and a north pole at others.

REASON: The bathtub is a huge piece of steel, and any large piece of steel will become magnetized by the magnetism of the earth. If the tub is in a north-south direction in the room, one end will be north and the other south to the compass.

This applies to the Southern states. In states farther north, where the earth's magnetic lines are more perpendicular, the north and south points of the tub are likely to be top and bottom.

Chapter 9

Air, Air Pressure & Gases

A BALLOON TRICK

NEEDED: A rubber balloon, a glass jar, water and soap, a pencil.

EXPERIMENT: Inflate the balloon so it is slightly larger than the mouth of the jar. Try to force it into the jar; it is difficult if not impossible.

Then, slide a pencil down beside the balloon, and the balloon may be pushed into the jar.

A Balloon Trick

REASON: The balloon seals the opening of the jar so air cannot escape around it, and as it is pushed, it tends to compress the air in the jar slightly. It cannot be easily pushed against the air pressure.

When the pencil is used, the trapped air can flow out of the jar through openings at the sides of the pencil, so there is no compression. It may be necessary to wet the balloon with soapy water so it will slide into the jar.

The Way the Wind Blows

THE WAY THE WIND BLOWS

NEEDED: Two rubber balloons and a means of connecting them together.

EXPERIMENT: Blow up both balloons, and connect them with a tube or spool (as in the drawing). The smaller balloon will shrink while the larger balloon gets larger.

COMMENT: This is the way some of the books tell it, but it does not always work out this way. It does work with two soap bubbles joined by a tube so the smaller bubble feeds air into the larger one. This is because of the difference in surface tension, which is greater in the smaller bubble.

There cannot be a balance in the pressure exerted by two bubble films. There usually is such a pressure balance in the balloons because of the strength characteristics of the balloon surfaces.

POURING COLD AIR

NEEDED: A cardboard box or styrofoam ice chest, a refrigerator, a warm room or warm day, without wind.

EXPERIMENT: Hold the container at the open refrigerator door for 30 seconds. Place the lid on it carefully, and move it to a

Pouring Cold Air

warm place where the air is still. Hold it above the face, open the lid, and the cold air will pour down on the face.

REASON: Cool air is a little more dense and heavy than warm air, and so it will pour out of the refrigerator into the container. In the same way it will pour out again, from the container.

If a cardboard box is used it may be closed with a folded newspaper.

The Arrow

THE ARROW

NEEDED: A needle and short thread.

EXPERIMENT: Throw the needle with the point first, at a window curtain. It is likely to go straight as an arrow, and stick into the curtain. Try throwing it without the thread, and it is not likely to stick into the curtain, but will fall to the floor.

REASON: When the needle is thrown it is given momentum. The thread tends to hold it back, because of air resistance, and this slight pull backward on the eye end of the needle tends to keep the point moving forward. This is the principle of the arrow. Without the thread the needle direction cannot be controlled well. This control by the thread is similar to the balance stick used to guide early rockets.

Wind Resistance

WIND RESISTANCE

NEEDED: Skates and two large cards.

EXPERIMENT 1: Hold the cards as shown and they act as brakes.

EXPERIMENT 2: Hold the cards edge forward, and they will cut through the air easily; not much air has to be moved to allow them to move through it.

EXPERIMENT 3: Hold the cards as shown in the drawing and wave them back and forth. It may be possible to stand up and allow the motion of the cards to move *you* back and forth, an illustration of Newton's third law of motion. (For every action there is an equal and opposite reaction.)

THE GLIDING GLASS

NEEDED: A plastic cup or glass, hot water, a smooth surface such as a plastic counter top or varnished table top.

The Gliding Glass

EXPERIMENT: Rinse the glass with hot water, leave a little water in it, and invert it on the smooth surface. It will "skitter around" as if on ice, with almost no friction. It can be moved with a breath or a feather.

REASON: As the water is poured out of the glass it is replaced by room air. Heat stored in the glass and water heat the air somewhat, it expands, and its pressure lifts the glass a tiny distance from the surface of the table. The glass floats on a film of water and a cushion of air; it may not touch the surface on which it rests at all.

COMMENT: This is the same principle used by "surface-effect-vehicles" or "hovercraft." The English have such a craft to cross the English Channel, from Dover to Calais, replacing the Channel-crossing ships which routinely made most of the passengers seasick.

Such a hovercraft has a fan which blows air downward into a very large "shroud ring." Although the discharge pressure of the fan is not very high, the total force exerted (pressure times area) is enough to let the hovercraft hover.

A BERNOULLI TRICK

NEEDED: Two empty cans on a smooth table.

EXPERIMENT: Place the cans as shown, blow down between them, and they will tend to roll together.

REASON: Bernoulli discovered that air in motion exerts less pressure than still air. In this case the moving air between the cans reduces the pressure there somewhat, so the greater pressure of the atmosphere on the opposite sides of the can can push against them.

The breath must be blown forcefully to make this trick work. The cans must be close to one another at the start of the experiment.

A Bernoulli Trick

HAIR DRYER

WIND DIRECTION

TOP BOARD

SAIL

BLOCK AND SAIL MOVE THIS WAY

BOTTOM BOARD

Sailing

SAILING

NEEDED: A cart made to roll easily on dowels, a sail mounted on a block of wood resting on the cart, a fan or hair dryer.

EXPERIMENT: Set up the demonstration as shown in the diagram, and by directing the wind as shown, the cart may be made to "tack" into the wind. This means that the cart may be made to travel in a direction which is upwind. This is a principle of sailing.

COMMENT: The writer used two pieces of wood, 1 by 4 inches, with straight dowels between to make the upper board move with very little resistance. The block holding the sail must be movable so the sail can be adjusted to the wind direction.

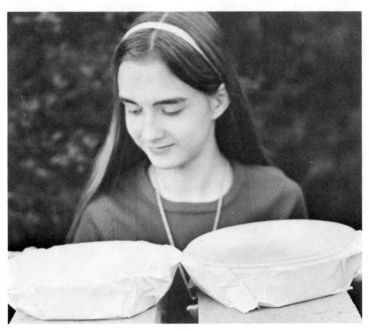

Air Pollution

AIR POLLUTION

NEEDED: A white pan, or any pan with a white paper in the bottom.

EXPERIMENT 1: Fill the pan with water, set it in the open where it will not be disturbed. Look at it after two or three days.

EXPERIMENT 2: Try two pans, one filled with water and the other dry.

OBSERVATION: Air pollution is often thought of as gases and mixtures of gaseous products, but air always contains solid matter also. It is the solid or "particulate" matter that can be caught and seen in a pan.

In a city the particulate matter from smokestacks can be amazingly heavy. As much as five tons of it can fall on every square mile of a city within 24 hours.

An Erratic Ball

AN ERRATIC BALL

NEEDED: A table tennis ball, a thread or light string, sticky tape, a soda straw with a flexible section.

EXPERIMENT: Attach the ball to the end of the string with a tiny piece of tape. Suspend the ball by the string. Blow upward against the ball, through the straw. Blow slowly, then harder.

REASON: Blow gently up against the ball, just off center, and the ball will try to "outflank" the air jet by going around it to where it can again hang vertically. The ball's motion will be quite erratic.

Blow harder, and the ball will go into the center of the air stream and tend to remain there, even though the stream be moved aside a considerable distance from the normal center of the ball.

In this case the Bernoulli rule will apply: moving air exerts less pressure than still air, and so the ball tends to remain so that the pull of the moving air around it is nearly equal on all sides.

CRUSH THE JUG

NEEDED: A plastic jug, a hose attached to it, water.

EXPERIMENT 1: Fill the jug with water, and put a bath spray hose over the mouth. Invert the jug, and as the water flows out through the hose the jug will collapse.

EXPERIMENT 2: Another way to crush the jug with atmospheric pressure: simply suck air out of it with the mouth!

REASON: As water flows out of the jug, pressure above the water level is lessened, so that the greater pressure of the atmosphere crushes the jug.

This is an easy variation of the favorite laboratory demonstration of "crushing the can" with atmospheric pressure. Using the plastic jug instead of a tin can is easier, and the jug may be blown back into shape and used over and over.

If a bath spray hose is not available, the hose may be sealed into the cap of the jug with wax, or may be attached to a metal tube which may be sealed into the cap with wax.

Crush the Jug

BOUNCE A BALL

NEEDED: A ball such as a basketball or beach ball.

EXPERIMENT: Bounce the ball when it is inflated tightly.

Let out most of the air, so the ball is soft, and bounce it again. It will bounce very little.

REASON: As the ball strikes the floor air molecules exert their pressure against the floor through the fabric of the ball. The tightly compressed air in the ball has many times more air molecules exerting this pressure.

Each molecule is in rapid motion and exerts its own pressure. The more molecules the more pressure is exerted. A tremendous pressure is exerted on the body by the moving molecules of air, but this is not felt because there is equal pressure from inside the body.

Air is elastic, that is, it can be compressed then can resume its former volume. A ball bounces as the air expands again after being compressed by the blow against the floor.

Elastic Air

ELASTIC AIR

NEEDED: A soft drink bottle with a screw cap, a large nail, hammer, water.

EXPERIMENT: Drive the nail through the cap to make a hole (while the cap is not on the bottle). Put the cap on. Fill the bottle half full of water, turn it upside down, and the water will not pour out.

Hold the bottle upside down and blow into it. Take the mouth away, and water squirts out.

REASON: Surface tension across the hole acts like a sheet of rubber to keep the air out of the bottle and the water in. But when air is blown into the bottle through the hole, it compresses the air in the bottle above the liquid. Then when the pressure is removed, the air in the bottle expands to force out some of the water.

A fluid (air is a fluid) which tends to assume its former shape (or volume) when squeezed, twisted, or otherwise deformed is elastic.

The person blowing into the bottle may get a little wet. Don't ruin someone's clothes with this trick.

A Strong Little Finger

A STRONG LITTLE FINGER

NEEDED: A plumber's suction cup, a smooth surface such as a plastic table top, two people.

EXPERIMENT: Make a small hole in the suction cup. Moisten the cup and the surface on which it is to be placed. Have a strong boy press the cup down on the smooth surface, and pull it away. This can be done easily.

When a small child holds a finger over the hole on the next try,

143

the boy will find it difficult if not impossible to pull the cup from the surface.

REASON: If air cannot get in through the hole or around the edge of the cup, the rubber will stick very tightly to the smooth surface. The atmosphere pushes down at about 14.7 pounds per square inch, and since the mouth of the force cup could enclose an area of 16 square inches, the force required to pull it loose from the smooth surface could be more than 200 pounds. This represents the pressure of the atmosphere on the top of the rubber cup.

EASY CRUSHED CAN

NEEDED: A gallon-size plastic jug with a screw-on lid, boiling water.

EXPERIMENT: Put boiling water into the jug (do not try to boil water in the jug) and shake it with the lid *loose*. When steam and water have stopped coming out, screw the lid on tightly. The jug will begin to collapse. The action can be speeded up by running cold water on the jug.

REASON: As the steam in the jug condenses the pressure in the jug diminishes. Atmospheric pressure crushes it.

This has formerly been done with a metal can, which is crushed with equal ease by the pressure of the atmosphere. But by using the plastic jug a can is not ruined each time. To restore the jug to its former shape, put boiling water into it, screw the lid on tightly, then shake. Or, blow into it.

AIR EXPANSION

NEEDED: A fruit jar with tightly-fitting lid, a balloon, a pan of water, a stove.

EXPERIMENT: Place a few tablespoonfuls of water in the jar, place the jar in the pan of water, and bring it to a boil. Put some air into the balloon, tie it tightly, drop it into the jar, tighten the lid on the jar, and let the jar cool quickly. The balloon will expand and perhaps fill the jar.

REASON: The steam had driven most of the air from the jar, and when it cooled and changed back into water by condensation, it occupied very little space; therefore the air in the balloon expanded to fill some of the space. A gas (air) will fill any space in which it is confined.

AIR PRESSURE TRICK

NEEDED: A milk jug, a plastic bag, a coin, some string, some rubber bands.

144

Air Pressure Trick

EXPERIMENT: Tie the coin on the bottom of the bag as shown, so the string will not pull off of the bag. Place the bag in the jar, blow into the bag to inflate it, place rubber bands around the neck of the jug to fasten the bag, then lift the string. The entire jug will be lifted; the plastic cannot be pulled out of the jug without tearing.

REASON: As the bottom of the bag is pulled upward by the string, the air pressure between it and the bottom of the jug is reduced. The greater pressure of the atmosphere pressing down on the bag holds the bag in the jug, and the jug may be lifted.

AN AIR CAN

NEEDED: An old long-playing record, a wooden thread spool, a candle, a large rubber balloon, a smooth surface.

EXPERIMENT: Whittle one end of the spool down so that the balloon can be slipped over it. Attach the other end of the spool to the center of the record with candle wax. The holes in the spool and the record should match.

Inflate the balloon, slip its mouth over the spool, place the record on the smooth surface, release the balloon, and the record will glide with very little friction over the surface.

REASON: When the record rests on the surface it tends to remain there because of the friction created when the surfaces move

145

An Air Car

against each other. The air stream from the balloon puts a thin layer of air between the surfaces, eliminating most of the friction.

THE STUBBORN BALLOON

NEEDED: Two cans, a rubber balloon, soap suds, nail and hammer.

EXPERIMENT: Make a nail hole in one can. Put air into the balloon, and try to push it into the can without the hole. It may be impossible. Now try pushing it into the can with the hole. Place the balloon in the end of the can with the hole, suck air out of the hole, and see if the balloon can be drawn into the can that way. It is easier if the balloon is covered with suds.

REASON: To push the balloon into the can, friction between the metal and rubber must be overcome. In the can without the hole, air trapped in the can must be compressed also. As the balloon is pushed in farther the air pressure increases.

If air is removed from the can by suction the balloon may be drawn in, pushed actually by the greater pressure of the air outside the can.

FOUNTAIN IN A JUG

NEEDED: Gallon jug, coffee can, soda straw, a candle, hot water, ice water.

EXPERIMENT: Make a hole in the jug cap, insert the straw, and pour melted candle wax around the straw to seal it in the cap. Put hot water into the jug several times, to heat it thoroughly. Pour out the water, screw the cap on tightly, put the jug upside down in

146

Fountain in a Jug

the coffee can of ice water, and watch a fountain flow from the upper end of the straw.

REASON: As the air in the hot jug cools, its pressure is reduced, so that atmospheric pressure on the surface of the ice water can force some of it up the straw. As it comes out of the end of the straw, it cools the air in the jug more quickly, and the pressure is further reduced, so that water flows out faster. (Look up Charles' Law.)

If there is steam in the jug some of it is condensed by the cold water, and this adds to the vacuum effect.

Carbonated Beverage

CARBONATED BEVERAGE

NEEDED: Two tall glasses, with a fresh soft drink in one and plain water in the other; a small cage made of screen wire, some insects, matches, covers (cardboard is good) for the glasses.

EXPERIMENT: Lower the cage of insects into the glass above the water, the insects are not affected. Lower the cage into the soft drink glass, keeping it above the surface of the liquid, and the insects will soon die. A lighted match will continue to burn above the plain water, but will be extinguished if lowered above the soft drink.

REASON: "Carbonation" of a beverage means that it contains dissolved carbon dioxide gas. Some of the gas constantly escapes, filling the space above the liquid, pushing out the air. The insects die, or the match goes out, for lack of oxygen in the air.

This does not mean that carbonation of a beverage makes it unhealthful. What we drink goes to the stomach; what we breathe goes to the lungs. Our lungs, not our stomachs, require oxygen.

THE BALLOON FALLACY

NEEDED: A yardstick, balloon, paper cup, string, salt, a pin.

EXPERIMENT: Suspend the stick at the middle. Tie a cup at one end and a blown-up balloon at the other. Pour salt into the cup until the stick is balanced.

Let the motion stop, then stick the pin into the balloon. The cup with salt is now heavier. (If pieces of balloon fall to the floor, put them back on the string so the loss of weight cannot be attributed to loss of rubber.)

REASON: This experiment is often given to prove "air has weight." The air in the balloon has weight, of course, but the air outside the balloon has weight, too, and tends to buoy it up. If the pressure inside the balloon could be exactly that of the air outside it, there would be no difference after the balloon is punctured.

What this does prove is that *compressed* air inside the inflated balloon is heavier than the air outside the balloon, which has less pressure.

Another way to try this is to balance two balloons on the stick, and puncture one of them.

ANOTHER MYTH EXPLODED

NEEDED: A bowl, a jar, a lighted candle, water.

EXPERIMENT: Drop wax on the bottom of the bowl to attach the candle. Pour water around the candle, light it, place the jar over it. Air bubbles out at the bottom of the jar, as the air in the jar is heated and expanded by the hot flame. Soon most of the oxygen

Another Myth Exploded

in the jar is used up, and the candle goes out. Then, as the air in the jar cools and contracts, water is forced up into the jar by pressure of the atmosphere on the surface of the water.

REASON: This proves the air is 1/5 oxygen, since water comes up to fill about 1/5 of the jar. *This is not the correct explanation.* The water comes up into the jar mainly because of the expansion and contraction of the air in the jar.

However, the oxygen is mostly consumed by the flame, and about an equal volume of carbon dioxide is produced. Carbon dioxide is about 26 times as soluble in water as the oxygen, and so, some of the carbon dioxide is dissolved in the water, adding somewhat to the lessening of the air pressure in the jar.

A Bad Barometer

A BAD BAROMETER

NEEDED: A quart milk bottle, a balloon, soda straw, a box, a wax candle, rubber bands to fasten the balloon.

EXPERIMENT: Fasten a section of balloon rubber over the mouth of the bottle with the tight rubber bands, and attach the end of the straw to the center of the rubber with melted wax. Mark the height of the other end of the straw on the box, and as the barometric pressure varies, the straw will move very slowly up and down.

REASON: This is often suggested as a simple school science project. As a barometer it cannot work, because variations of temperature move the straw more than barometric pressure. As temperature rises, air in the bottle expands, pushing the rubber diaphragm up and the pointer down. No commercial barometer is made exactly like this.

THE ERRATIC BALLOON

NEEDED: A helium-filled balloon and an automobile.

EXPERIMENT: Hold the balloon by the string in a closed automobile. Let someone drive the vehicle in a circle. All objects in the car will tend to move toward the outside of the circle—except the balloon. It will float in the opposite direction.

REASON: The balloon tends to rise because its helium is lighter than air. As the automobile turns, the heavier air tends to move toward the outside of the circle, as do the people and other objects. The helium, being lighter, tends to move in the opposite direction.

AN ANALOGY: Float a ping-pong ball on water in a fish bowl, and turn around with the bowl in a hand. The water will move toward the outside of your circle, and the ball will move toward you.

THE GHOST FLAME

NEEDED: A candle or gas flame, piece of screen wire, a pair of pliers, a match.

EXPERIMENT: Hold the screen as in the left-hand drawing, so that half of the flame is cut off. There is no flame above the screen. Light the place above the screen where there is no flame, and a ghostly flame appears.

REASON: When the screen is held midway of the flame it conducts so much heat away that the flame ends there. The gases do not get hot enough to ignite, but they are there, passing through the screen mesh. When lighted they will burn. If a gas flame is used, the

The Ghost Flame

gas may be lighted above the screen, and it will not catch fire below it, for the same reason.

SMOKE RING CANNON

NEEDED: A coffee can with two plastic covers, incense or something else to produce smoke.

EXPERIMENT: Cut the bottom of the can out. Place plastic lids over both ends of the can; one lid should have a hole in it. The hole can be the size of a dime if the can is small and the size of a quarter if it is a large one. Put smoke into the can, and rings are produced by tapping the lid opposite the hole.

Smoke Ring Cannon

REASON: When the lid is tapped, an air wave comes out at the hole. The wave is similar to a low frequency sound wave.

A little smoke comes out with each wave. The edges of the hole drag on the edges of the disc of smoke and air, so that a toroidal or "doughnut" circulation is begun. This shape continues as the rings move away slowly and beautifully. See also "Smoke Rings in Water."

Chapter 10

Heat

ICY TEMPERATURE

NEEDED: Paper cup, refrigerator or freezer, thermometer, water.

EXPERIMENT 1: Place the thermometer in the cup of water,

Icy Temperature

let it stay in the freezer overnight. Check the temperature. It is likely to be considerably below the freezing point of water, 32 degrees Fahrenheit or 0 degrees Celsius. The temperature should go to 32 degrees when the ice begins to melt, and should remain at that temperature as long as the thermometer bulb is encased in ice.

EXPERIMENT 2: If an oven thermometer is available an experiment on the temperature of heated water is possible. Heat some—the temperature rises until it is about 100 degrees Celsius (212 degrees Fahrenheit). This varies with atmospheric pressure. The water temperature remains at that level; additional heat applied is used in the boiling.

Water under pressure can be at a much higher temperature, however.

REASON: Ice can be much lower in temperature than its freezing point, but should not go higher than that as long as it remains ice. Additional heat absorbed by it will be expended in the melting process after it reaches 32 degrees.

Dark Heat

DARK HEAT

NEEDED: Two like metal cans, dull black paint, two thermometers, water, sunlight coming through a window.

EXPERIMENT 1: Spray one can black inside and out. Fill

both cans with tap water, an equal amount in each. Place the thermometers in the cans, and put them side by side in the sun. Take readings every ten minutes, and it will be found that the water in the black can warms more rapidly.

REASON: Much of the light that falls on the bright metal can is reflected away, while more of the light falling on the black can is absorbed by the black paint and changed into heat.

EXPERIMENT 2: Put warm water in both cans, and set the cans on a table, being careful not to use water hot enough to break the thermometers. The black can will radiate heat away more quickly and will cool faster.

A Water Dance

A WATER DANCE

NEEDED: A skillet, heat, water.

EXPERIMENT: Heat the skillet so that when a drop of water falls on it the drop will dance around as it evaporates. See how long it takes the drop to evaporate. Let the skillet cool some, drop water on it, and the drop will evaporate faster if it does not dance, even though the skillet is cooler.

REASON: As the drop touches the hottest skillet a little steam is formed, under the drop, enough to hold the bulk of the drop above the hot surface, in effect, insulating it somewhat so heat travels into it comparatively slowly.

If the drop does not dance, it touches the hot metal over a large area of its surface, and so absorbs heat faster, and evaporates faster.

155

Freeze to Finger

FREEZE TO FINGER

NEEDED: An ice tray just out of the freezer.

EXPERIMENT: Take the cubes out. Touch the finger to one, and it is likely to freeze to the finger. Try wetting the finger before touching the cube, and see if it sticks more tightly to the finger.

REASON: As the finger approaches the cube, the warmth of the finger melts a tiny bit of the cube. As heat is conducted away from the contact through the cold cube, the tiny bit of water freezes into ice.

It is possible to have a hand frozen to a very cold piece of metal so that skin is pulled off the hand. The ice cube stunt is safe, however, because the finger continues to get a little warmer at the contact point all the time and will again melt the ice at the contact point.

A RUBBER BAND MYSTERY

NEEDED: A weight suspended from a rubber band, a match.

EXPERIMENT: Hold the flame of the match near the rubber band, and as the band is heated, it contracts, lifting the weight.

REASON: While most substances expand when heated, some will contract. Eleven volumes of ice, when heated, make only about ten volumes of water, for example. Such substances as rubber contract because of their increased stiffness when heated; therefore, more force is required to stretch the rubber band. For better technical understanding, look up Young's modulus.

Another Rubber Band Trick

ANOTHER RUBBER BAND TRICK

NEEDED: A thick, wide rubber band or piece of inner tube rubber.

EXPERIMENT: Stretch the rubber band quickly while holding it to the lips. It will get warmer. Let it contract, and it cools.

REASON: Here's one of those questions that cannot be answered satisfactorily and simply. Gases, when compressed, get hot, and cool when allowed to expand. It is thought that the rubber molecules act in the same way, but this is not definite.

If you increase the temperature of a rubber band its tension increases. If you increase the tension of a rubber band (with all else constant) its temperature increases.

A FIREPROOF CLOTH

NEEDED: Two ounces of rubbing alcohol, one ounce of water, a square of cotton cloth.

EXPERIMENT: Mix the water and alcohol and soak the cloth in it, hold it at a distance with a coat hanger and try to light it with a match. It seems to burn, yet the cloth remains unharmed.

REASON: The alcohol burns when lighted with the match, but the heat produced is not sufficient to evaporate the water from the cloth, which keeps the temperature of the cloth so low that it does not burn.

Do not try to fireproof clothing in this manner.

Fireproof Cloth

FIREPROOF PAPER

NEEDED: Two pieces of paper, two matches, a drinking glass.

EXPERIMENT: See how easily the paper burns when lighted with a match. Wrap the other piece around the glass, so that it touches at all points, then try to light it. It will not burn.

REASON: The glass absorbs the heat from the match as it goes through the paper, so that the paper does not reach its kindling temperature. Also, the glass tends to cut off the oxygen of the air necessary for burning.

BURNING PAPER

NEEDED: A tightly rolled paper, a crumpled paper, a match.

EXPERIMENT: Try to light the rolled paper with a match, and it is very difficult. Light the crumpled paper, and it burns readily.

REASON: When heat is applied to the rolled paper it does not burn easily for two reasons. First, the air cannot get to the inside sheets, and they cannot burn without oxygen from the air. Second, when the paper is rolled, much of the heat from the match is conducted into the roll, and the outside layers of paper cannot get hot enough to ignite. The crumpled paper is heated quickly by the match, and since air is all around it, it burns easily.

How Hot the Water?

HOW HOT THE WATER?

NEEDED: Two paper cups, water, salt, a heat source.

EXPERIMENT: Boil water in the paper cup in the regular way. The paper does not burn. Heat dry salt in the other paper cup; the paper burns.

REASON: Water cannot be heated above its boiling point—any more heat applied is used up in boiling the water. Since the burning point of the paper is higher than that of boiling water, and since the paper conducts the heat readily into the water, the paper does not burn. The salt does not conduct the heat away from the paper readily, and has a melting point of about 800 degrees centigrade which is much higher than the burning point of the paper. Therefore, the paper is heated to its burning point.

THE STUCK JAR LID

EXPERIMENT: A jar lid that cannot normally be removed when cold can be made loose enough to be turned when heated under hot water.

REASON: First, the lid expands when heated, as most substances do, and may become looser for this reason. Second, if there is a rubber gasket in the lid, it will contract when heated, making the lid loose enough to be twisted off. Remember rubber is one of the few substances that will contract when heated.

Contract or Expand

CONTRACT OR EXPAND

NEEDED: A dish of water in the freezing compartment of a refrigerator, a dish of melted paraffin.

EXPERIMENT: Let the water freeze in the refrigerator, and let the paraffin freeze at room temperature. The water expands on freezing, and the paraffin contracts.

REASON: Water, type metal, and cast iron are among the very few substances that expand when they freeze. Most others continue to contract as their temperatures are lowered.

Water contracts until its temperature is lowered to four degrees centigrade, then it begins to expand. While freezing it expands nearly 10 percent, then, as its temperature is further lowered, it contracts again.

The expansion of water on freezing is of great importance to the life forms in lakes and ponds. If it would contract, as most substances do, the lake would freeze solid from the bottom upward, making it impossible for fish to live.

A WARM CUSHION

NEEDED: A kitchen grater, old Styrofoam containers, a cloth sack for making the cushion.

EXPERIMENT: Grate the styrofoam into small pieces, stuff them into the cloth, and sit on the cloth. The cushion seems to generate heat.

REASON: The Styrofoam is a good insulator for heat, so most of the body heat is retained rather than dissipated. It is a "heat trap."

WHY THE ALCOHOL RUB?

NEEDED: Rubbing alcohol and water.

EXPERIMENT: Extend your hands, and have someone pour water on one and alcohol on the other. If both liquids are the same temperature the alcohol will feel much cooler to the hand.

REASON: Alcohol evaporates faster than water, and when a liquid evaporates it must take heat from its surroundings. Therefore, the alcohol takes more heat from the hand *faster* than does the water.

Nurses rub patients with alcohol because it is cooler, and it also has some germ-killing power.

REFLECTION OF HEAT AND COLD

NEEDED: A heat source such as a corn popper or hotplate, a pitcher of ice cubes, a curved piece of metal or cardboard, some kitchen foil, a book, two thermometers.

EXPERIMENT: Cover the curved card or metal (the author used the side of a coffee can cut with a can opener and tin snips) with crumpled foil to be used as a reflector. Arrange the heat source, the reflector, and a thermometer so that the heat reflecting from the foil will reach the thermometer. A book between the heat source and the thermometer will prevent the direct heat radiation from affecting the thermometer.

The heat will be reflected and the higher temperature on the thermometer will be easily read if the heat source, reflector, and thermometer are in the right positions. Remove the heat source and place the pitcher in its place. Use the other thermometer which has not been heated and see if it will show a drop in temperature from the reflected cold from the ice cubes.

REASON: There was no change in the temperature shown by the second thermometer. Radiant heat is energy and can be reflected. Cold is the relative absence of a certain amount or degree of heat, and cannot radiate anything.

Chapter 11

Light

STROBE-LIKE LIGHT

NEEDED: A fluorescent light.

EXPERIMENT 1: With a fluorescent light as illumination, move the hand up and down rapidly. More than four fingers and one thumb appear to be seen.

Strobe-Like Light

REASON: This is a case in which the fluorescent light is a stroboscope. Its light goes dim then bright again 60 times per second, reversing itself 120 times each second.

When the light is bright we see the fingers. Then while the light is dim the hand moves to another position and we see the hand again when the light brightens, but in a new position. Retention of the view in both one position and the other gives the appearance of more than four fingers and one thumb. This retention is a trick of the eyes.

EXPERIMENT 2: Hold the hand in front of the television screen. The same effect is seen, and for the same reason. The TV picture goes on and practically off rapidly.

EXPERIMENT 3: Try this in front of a home movie projector while it is operating. Here the picture goes entirely off then on again rapidly, perhaps 24 times a second.

Fluorescent Light

FLUORESCENT LIGHT

NEEDED: A fluorescent light and an electric fan.

EXPERIMENT 1: Watch the fan as it speeds up when turned on and as it slows down when turned off. It seems to go forward then backward and vice versa.

REASON: A fluorescent lamp does not burn steadily, but gets bright then dimmer 60 times a second as the current in the line reverses itself 120 times a second. The illusion is a stroboscopic effect and is more noticeable in some of the older lamps.

If the fan turns at such a speed that the blades are always in one position when the lamp is brightest then the fan seems to be stopped. If the fan turns a little more rapidly the blades seem to turn forward slowly. If the blades turn a little slower they will seem to be turning backward slowly because of the position of the blades when the light from the lamp is at its brightest point.

The human eye does not notice the fluctuation in the brightness of the fluorescent lamp.

EXPERIMENT 2: Run the fan in a room where the TV set is on. Turn out all other lights. The same effect is noticed and for the same reason. Also, the fan in the room where motion pictures are shown will produce the stroboscopic effect.

Pin-Hole Vision

PIN-HOLE VISION

NEEDED: Cardboard, a pin, a stick four or five inches long.

EXPERIMENT 1: Hold the stick (the author used a Popsicle stick) against a printed page, the other end against the forehead.

Close one eye and try to read the fine print. It may be impossible. Hold the card so you look through a pinhole in the card with one eye. The print may be clear.

REASON: The eye cannot normally focus at such a short distance. But when the light reaching the eye comes through a tiny hole, focusing of the eye lens is not necessary. This is because the pinhole allows most of the light to pass through the middle portion of the eye directly to the center of the retina which is most sensitive. No refraction is required.

The diagrams should help to explain.

EXPERIMENT 2: Try pinhole vision if glasses are needed for clear distant seeing. Look through the pinhole—without the glasses.

Light Waves

LIGHT WAVES

NEEDED: Screen wire, a candle or bare electric bulb in a dark room.

EXPERIMENT: Look at the light through the screen. Streamers of light will be seen at right angles crossing at the light source. The lamp must be across the room rather than close to the screen for this effect.

REASON: Each little hole in the screen acts as if it were a source of light, and the light waves come through in step (in phase). But when they come together in the eye their different path lengths vary and they are not all in phase. They act to cancel each other out or reinforce each other adding to their brightness. This is called an interference phenomenon. Look up Fresnel's rule.

Try looking at the light through cloth, and also through two pieces of screen wire. Turn one of the pieces of screen as you look.

A Rainbow Puddle

A RAINBOW PUDDLE

NEEDED: Dark colored bowl, water, oil.

EXPERIMENT: Moisten a toothpick or match stem with oil, and touch it to the middle of the water surface in the bowl. Look at the surface from various angles. Beautiful rainbow colors will be seen, as often seen in a puddle of water on a street.

REASON: Light rays striking the water surface are reflected in two ways: some from the surface of the oil and others from the surface of the water under the oil. The distance between the oil and water surfaces is such that "interference" in the two rays results.

The interference serves to cut out some colors, allowing others to be seen.

As the oil film moves, its slight wedge shape allows different colors to show at different parts of the film. See diagram.

The author found that Three-In-One Oil gave best results in this experiment, although some other oils may be used.

Mystery Colors

MYSTERY COLORS

NEEDED: An electric fan and a fluorescent light.

EXPERIMENT: Hold the fan so the fluorescent tube is reflected from one of the blades. Start the fan and watch carefully, colors may be seen in the reflection.

REASON: Phosphors inside the fluorescent tubes glow in red, green, and blue, but the colors are so mixed that the result is white light. Each of the phosphors has its own time of buildup of brightness and decay, and when the fan blades are turned at a specific speed they "tune in" to the time of one of the phosphors. At another speed they "tune in" to the time of another.

This effect is not easy to see. Only at certain definite speeds are the different colors seen; the colors change as the fan slows or speeds up. The strobe effect also is seen in the fan blades; they appear to stop and reverse themselves as the speed of the fan changes.

SUN'S RAYS

NEEDED: Observation.

OBSERVATION 1: The sun, on a foggy morning, shines through the trees in a very beautiful pattern as shown in the photo. Sometimes it makes a similar pattern shining through the clouds, and the old-timers explained this as "the sun drawing water."

REASON: The phenomenon occurs as the sun shines into and through tiny droplets of water (in the sky the water may be frozen). The light reflects and refracts in almost every direction. Some light from each droplet may reach the eye.

Sunlight shining on water droplets from a different angle makes the rainbow.

OBSERVATION 2: Note that the rays radiate apparently from a point just beyond one of the trees and extend from that point in all directions. We would expect the rays to seem parallel. Is this a matter of geometry?

THE TELESCOPE

NEEDED: Two magnifying glasses, one larger than the other.

Sun's Rays

The Telescope

EXPERIMENT: Hold the large glass at arm's length, and adjust the distance of the other from the eye so that a distant object, seen through both lenses, seems close.

REASON: This shows the principle of the refracting telescope. The objective lens forms an inverted real image of the object looked at. The eye lens views the real image of the objective lens and shows it to the eye as an enlarged and inverted virtual image.

A real image can be shown on a screen. A virtual image cannot. Reflection in a mirror is a virtual image; you cannot touch it or put it on a screen. Your face on the movie screen when your vacation movies are shown is a real image.

Most large telescopes are the reflecting type mainly because lenses cannot be made as large as mirrors. Purpose of the large size of the mirror or objective lens is not to magnify, but to gather light. The larger it is the more light it gathers.

TRICKY COLORS

NEEDED: Red, blue, or green cellophane, crayons the same colors, white paper. Cellophane must be transparent.

EXPERIMENT: Draw something on the white paper using the red and either blue or green crayons in the same drawing. Look at the drawing through the red cellophane, and the red lines almost disappear; look at it through the other cellophane, and only the red lines show distinctly.

REASON: Light from the sun or a lamp contains the primary colors, and so the red lines can reflect red from the original light, and the blue or green can reflect their colors. White light from the paper is reflected, but as it passes through the red cellophane nearly all of it is absorbed except the red. This makes the white paper and red lines appear near the same color of red.

It is the same for the other colors. The blue and white reflect but all the colors in the white are absorbed by the blue cellophane except the blue. Practically no blue is reflected from the red lines, and so they appear black. When the drawing is seen through the red cellophane the blue or green lines appear black because they do not reflect red.

Reflection Under Water

MIRROR-LIKE SURFACE OF WATER

REFLECTION UNDER WATER

NEEDED: A jar of water and a lighted candle.

EXPERIMENT: Place the candle behind the water jar as shown in the diagram. Look upward at the water surface, and the candle will be seen reflected, upside down.

REASON: We are accustomed to seeing trees and cliffs reflected from the upper surfaces of calm lakes. The under surface of still water reflects in the same way. Skin divers can see this phenomenon if the surface of the water in which they are swimming is sufficiently calm.

The magazine *Popular Science* suggested this phenomenon as a possible explanation for some of the sightings called unidentified flying objects. Automobile or other lights can reflect from the bottom of an atmospheric "layer" to viewers at a distant place.

RAINBOWS IN A BUBBLE

NEEDED: Bubble blowing equipment.

EXPERIMENT: Light is reflected from both front and back sides of the film that makes the bubble. Light reflected from the inside surface reaches the eye a little later than that reflected from the outside surface, because it has a slightly longer path.

Because of what is called "interference" some of the rays reach the eye in phase, and are bright, while others are out of phase and tend to cancel each other out. Varying thickness of the bubble film accounts for different colors. When the thickness is very small the waves all cancel each other, and the appearance is dark.

SODIUM LIGHT

NEEDED: Alcohol, salt, a jar lid, a match, a large pan or board, different colored objects.

EXPERIMENT: Put some salt into the lid, then pour alcohol into it. Light the alcohol, turn out the lights so the room is dark, then sprinkle salt down into the flame. Look at the different objects.

REASON: Rubbing alcohol is not good for this. The author tried methyl alcohol (wood alcohol), and a red box looked black. Fuel alcohol works, and its flame made the red box show as yellow. The difference is in the chemicals in the alcohol itself which add their colors to the flame. A pure sodium flame, from the sodium in the salt, would be pure yellow with no other colors.

DRIVING IN THE FOG

EXPERIMENT: When headlights are turned to the bright or "up" position it is more difficult to see the road through the fog.

REASON: When a light ray hits a fog droplet, the light is reflected and refracted in all directions, so that the droplet would have the appearance as shown in the bottom right drawing.

The drawing of an enlarged fog droplet, at bottom left, shows some of the ways in which the light rays can reflect and refract when they hit the droplet. The same ray is reflected and refracted many times, as it hits many droplets, giving an overall dispersion or scattering of the light. When the automobile beams are turned down the light hits fewer droplets, so that the fog is less bright and the road is seen more clearly.

RAINBOWS IN A RECORD

NEEDED: One long-playing phonograph record.

EXPERIMENT: Hold the record so that light coming toward you shines on the record grooves. Rainbow colors may be seen.

REASON: The light is reflected from the irregular surfaces of the needle grooves, causing interference of the reflected light. This phenomenon is similar to that seen when light is reflected in rainbow colors from a soap bubble. The bubble rainbow was explained in a previous experiment.

A CANDLE FLAME SHADOW

NEEDED: A candle, a white or gray card, sunlight.

EXPERIMENT: Hold the candle so that its shadow is cast on the card. The shadow of the flame may be seen, but rather faintly.

REASON: The flame is composed largely of carbon particles in varying stages of burning. The particles do not look black in the yellow flame, but they are there and can make a faint shadow.

Also, what we see as a shadow is partly the refraction of the

A Candle Flame Shadow

light as it passes from cool air through the hot flame gases and back again into cool air. Light is refracted when it passes from one medium to another of different density, and the hot gases of the flame are less dense than the cool surrounding air.

Chapter 12

Household Hints

A WEED GOOD TO EAT

NEEDED: A lawn not too well kept.

EXPERIMENT: Look for the broad-leafed plantain weed, "plantago." It grows flat against the ground, its leaves growing directly from the root. Try to isolate one of the plants so it can grow a slender spike which will blossom into greenish flowers and later produce seeds. This life cycle is seldom noticed.

Many weeds and shrubs are poisonous. But plantain is good to eat, especially when the leaves are young and tender. The older leaves may taste bitter. The leaves may be boiled and served with butter and a little vinegar, or may be used in salads. They are good rabbit food.

MAKE SOUR MILK

NEEDED: Vinegar, fresh or evaporated milk.

EXPERIMENT: When sour milk or buttermilk is needed in a recipe, and there is none, put one tablespoon of vinegar in one cup of fresh milk, and stir. It may be used as sour milk. Diluted evaporated milk may be soured in the same way.

REASON: Vinegar is about four percent acetic acid. Sour milk contains lactic acid, formed from lactose sugar (present in sweet milk) by action of bacterial enzymes. Either acid serves the same purpose in the cooking recipe, and both are desirable in food in the small amounts involved. Lemon juice, which contains citric acid, may be used.

A Weed Good to Eat

Make Sour Milk

Chemistry and Curds

CHEMISTRY AND CURDS

NEEDED: Tomatoes to be cooked with milk or cream, soda.

EXPERIMENT: Cook tomatoes with milk or cream, and curdling will likely result. Try cooking them with milk or cream to which a pinch of soda has been added, and—no curdling.

REASON: The curdling is due to the acid in the tomato juice causing the protein of the milk to coagulate. Baking soda neutralizes the acid in the tomato juice and the protein does not coagulate in the absence of acid.

Fresh milk does not coagulate, but sour (acid) milk which results when lactose of the milk changes into lactic acid does coagulate.

Brown Cupcakes

BROWN CUPCAKES

NEEDED: Observation as cupcakes are baked in vari-colored paper cups.

OBSERVATION: The cakes baked in darker colored paper are browner.

EXPERIMENT 1: Dark clothes are warmer in winter—not quite an exact statement. Put on a light colored glove and a dark glove. In sunlight the dark glove will feel warmer; it absorbs more energy. But in darkness it will be colder; it emits more energy. It does not have to be winter.

EXPERIMENT 2: Dark colored paper or cloth will melt through snow faster than light colored paper or cloth—as long as light energy is being absorbed by them. In darkness the dark colors will radiate or emit more energy and be colder.

EXPERIMENT 3: Note that ice will melt more quickly on dark pavement than on light pavement—in daytime when light shines. In the dark the darker pavement (with starting temperature the same as the light colored pavement) emits more radiation and can be colder.

REASON: Light colors reflect more light and heat than dark colors, and dark colors absorb more light, turning it into heat. Therefore, the cakes baked in dark paper get just a little hotter, and browner.

DOUBLE BOILING

NEEDED: A double boiler, a candy thermometer, water, a stove.

EXPERIMENT: Get the double boiler steaming, and take the temperatures in upper and lower vessels. The author, using this crude method, obtained readings of 208 degrees in the lower vessel and 200 degrees in the upper vessel.

Add three heaping teaspoonfuls of salt to the lower vessel, let it steam, and take the temperatures again. The author's readings this time were 216 degrees in the lower vessel and 204 degrees in the upper.

REASON: Adding salt raised the boiling point of the water in the lower vessel, and at the same time the temperature was raised in the upper vessel. (Incidentally, salt added to water also lowers its freezing point.)

In the author's test four cups of water were used in the lower container and two in the upper container.

Double Boiling

GELATIN AND EGG

NEEDED: Gelatin, an egg

EXPERIMENT: Note that gelatin is a solid at room temperature and a liquid when heated. It turns back to solid when cooled. Remember that egg white is liquid at room temperature and becomes solid when heated. The egg remains solid when cool.

REASON: Egg coagulates when it is heated because small particles join one another to form a mass that remains together whether hot or cold. Gelatin is generally dissolved in a limited amount of water and heated mildly. When cooled this liquid may become a colloidal solid with the water as a part of the distributed total.

The water is not added to the egg but what is naturally present forms a permanent suspension when heated to the boiling point of water.

The clotting and setting of egg albumen is irreversible. The solidification of a gelatin solution is reversible.

Look up "colloid" and "suspension."

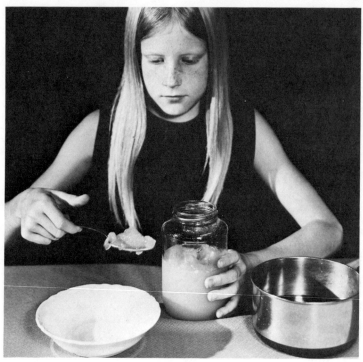
Gelatin and Egg

REMOVE HARDENED WAX

NEEDED: A deep freeze, or dry ice; blotting paper, a hot iron.

EXPERIMENT 1: When wax drops on a fabric and hardens, place the garment in the deep freeze for an hour or two. While it is still very cold, bend and break and rub the spot and the wax may turn to powder and brush off.

EXPERIMENT 2: If dry ice is available it is better because it is colder, but care must be taken in the handling of dry ice (solid carbon dioxide).

EXPERIMENT 3: If these methods do not work or are not practical, place the blotting paper over the spot of wax, and heat with the iron. Press firmly, and perhaps move the blotting paper around. The wax should be softened enough to flow up into the blotting paper, leaving the cloth.

COMMENT: The flow of the melted wax up into the fabric is by capillarity.

Remove Hardened Wax

Dry ice can harm the skin. Hold it with several layers of cloth or with tongs.

Dry ice will remove stuck-on chewing gum, even if in someone's hair!

HOW FRESH THE EGG?

NEEDED: Eggs.

EXPERIMENT 1: Put the eggs into a pan of water. If an egg is fresh it will rest on its side on the bottom of the pan. If it is a few days old one end will turn upward. If it is stale it will stand on one end, and if rotten it will float.

REASON: As the egg gets older some of the white and yolk lose water through the pores of the shell, and the liquid is replaced by air. The air accumulates in the large end of the egg, where there already was an air cell.

The air is buoyant and tends to lift the egg, more and more as the egg gets older.

EXPERIMENT 2: To tell a hard boiled egg from a raw egg: spin the egg on its side. The cooked egg will spin merrily, because the liquid inside the shell has been hardened and turns with the shell. When the raw egg is spun the liquid does not all spin with the

How Fresh the Egg?

shell, and actually acts as a brake to stop the spinning of the shell.

VITAMIN CHANGE

NEEDED: Orange juice, corn starch, tincture of iodine, a cooking pan and heat, water, glasses, spoons, eye dropper.

EXPERIMENT: Bring orange juice to a boil. Put uncooked juice into a container. Boil a half teaspoon of starch in half a glass of water, and put 20 drops of it into each of two glasses of water. Put two drops of iodine in each mixture, and stir. The color should be blue. Add 25 drops of heated juice to one glass, and stir. Then see how many drops of the other juice are necessary to produce the same color, in the other glass.

REASON: A mysterious combination of starch and iodine produces the blue color, and vitamin C destroys the color. This experiment should show that heating the juice destroys some of the Vitamin C. Less unboiled juice should be needed to remove the color.

It is not customary to boil orange juice, but for the purposes of this experiment the juice is easier to handle than apples or peas or other foods rich in vitamin C.

Vitamin Change

Boiling a solution of vitamin C destroys it. It is not the heat that does the job, however; it is the oxygen of the air that destroys the vitamin C, and boiling simply speeds up the action.

Chapter 13

Chemistry

SOLUTIONS

NEEDED: Observation.

EXPERIMENT: Note that sugar dissolves more easily in hot tea than in iced tea. Note that a cold soft drink is full of gas, but when

Solutions

it gets warm it gets "flat" or loses much of its gas.

COMMENT: Solutions composed of ordinary solvents such as water can usually hold more solid solutes, such as sugar or salt, when they are warm. The opposite is true if our solutes are gases. Solvents in general can hold more gases in solution when they are cold.

REASON: When a solid is dissolved in a liquid, the physical state of the solid may change to liquid, absorbing heat and lowering the temperature of the solution. Thus, solubility is increased as the temperature is raised.

In the case of gases in solution, raising the temperature increases the speed of the molecules of the gases and the solution, causing molecules of dissolved gas to leave the solvent.

Heating a solid is in the direction of melting it. Thus, it ought to "melt into" a liquid a littler easier.

Heating a liquid is in the direction of turning it to a gas. Thus, it ought to come out of the liquid a little easier, become a gas.

Solubility

SOLUBILITY

NEEDED: Household iodine, water, mineral oil, a jar with a tightly-fitting lid.

EXPERIMENT: Into a jar half filled with water put a few drops of iodine, enough to make a light-brown solution. Shake. Then put in two or three tablespoonfuls of mineral oil. Shake the mixture again, this time quite vigorously. Set the jar aside for a few minutes.

REASON: The color has left the water and has gone into the oil, showing that iodine is more soluble in mineral oil than in water. (Vegetable oil does not work in this experiment.) This could be called solvent extraction.

Almost all substances are more soluble in some liquids than in others. The explanations of this can get quite technical and often controversial.

Grow Crystals

GROW CRYSTALS

NEEDED: Two ounces of alum from a grocery store, water, heat, string, a jar, a paper clip.

EXPERIMENT: Heat the alum in a half cupful of water until all the alum is dissolved. Pour the solution into the jar, and hang the paper clip in the solution. After a few hours crystals of alum will be seen growing on the clip.

REASON: When the alum is all dissolved we have a saturated solution, which means the water will not hold any more of the alum. As the water cools it cannot hold all that has been dissolved: it becomes a supersaturated solution and some of the alum leaves the water, forming itself into crystals. The "super" part crystallizes out.

If crystals begin to form in the bottom of the glass and not on the clip, touch the clip to some of the dry crystals then hang it back into the solution. Crystals should begin to form on it very soon. A book suggests trying this trick with table salt. The salt water solution is likely to start rusting the metals rather than forming crystals.

A Color Mystery

A COLOR MYSTERY

NEEDED: Water, corn starch, household iodine, a stew pan, heat.

EXPERIMENT: Fill a measuring cup with hot water, pour the water into the pan, and put a tablespoonful of starch into it. Boil a few minutes, stirring constantly. Let it cool, add a few drops of iodine, and a blue color results. Heat the mixture again, and the blue color disappears.

COMMENT: We give the name "starch iodine" to a certain union of molecules of iodine with soluble starch. At room temperature the mixture is bright blue. What happens to produce the blue color, and what happens to make the liquid become colorless when heated has not been clearly understood, but is becoming better understood as this book is written.

Ice and Salt

ICE AND SALT

NEEDED: Two containers, two thermometers, water, ice, salt.

EXPERIMENT: Put much ice and a little water in both containers. Place the thermometers in the containers and note that the temperature is the same in both. Pour salt into one container, and note that the temperature goes down.

REASON: Unless there is heat loss, the temperature of melting ice stays the same in plain water. But salt on the ice lowers its melting point, and its temperature goes lower than the normal 32 degrees. Almost any soluble impurity will lower the melting point of ice, but salt is the one used in home ice cream freezers.

Salt is the most common substance used to melt snow and ice on highways and streets.

An interesting sidelight is that Fahrenheit used a mixture of ice and hydrochloric acid to produce what he thought was the lowest temperature possible. He put it at zero degrees on his scale.

Make a Salt

MAKE A SALT

NEEDED: Wine vinegar, baking soda, a tablespoon.

EXPERIMENT: Put vinegar into the spoon, then sprinkle soda into the spoon. Bubbles of carbon dioxide, CO_2, will form. Use barely enough soda to stop the bubbling. You have used an acid, the vinegar, and a base, the soda, and produced a chemical reaction which produced a salt.

Taste it! Taste the vinegar and soda before performing the experiment; the vinegar is very sour, the soda is bitter. The salt formed is not bad, its taste is near neutral.

COMMENT: Any kind of vinegar may be used, but wine vinegar has a more pleasant taste. Any acid will react with any soluble base to form water and a salt.

COVER THE SCRATCHES

NEEDED: Two tin cans, a knife, a cloth, water.

EXPERIMENT: Remove the labels from the cans. Make scratches on one can; leave the other unscratched. Wrap damp

Cover the Scratches

cloths around both cans. In a few days rust will have formed in the scratches.

REASON: "Tin" cans are iron, with a thin coating of another metal, perhaps mostly tin, which does not rust. Scratching exposes the iron, which will oxidize in the presence of air and water. Aluminum cans cannot be used in this experiment.

We now see why it is good to "touch up" or cover the scratches on an automobile or bicycle with paint. The scratches expose the iron to moisture and air and invite rust which can spread under the paint.

SULFUR IN YOUR EGG

NEEDED: A boiled egg, an old silver-plated spoon, a stainless steel spoon.

EXPERIMENT: Place some egg white on one end of each spoon, and some egg yellow on the other ends (or other spoons). Leave overnight. Next morning the silver-plated spoon will have a dark spot where the egg touched; the stainless steel spoon will be unchanged.

REASON: Eggs contain sulfur which unites with the silver to produce silver sulphide, a black tarnish. The sulfur does not readily unite with any chemical contained in the stainless steel spoon.

Try mustard instead of egg. It, too, contains sulfur in sufficient quantity to tarnish silver. The tarnish can be removed, but it is recommended that old spoons be used in this experiment. Dimes and quarters dated earlier than 1965 contained enough silver to use for this experiment.

AN EGGSPERIMENT

NEEDED: Vinegar, a jar, a raw unshelled egg.

EXPERIMENT: Place the egg in the jar, and fill the jar with vinegar. Watch. Bubbles begin to form on the egg, and it will rise to the top. Perhaps it will go down and up again. Leave overnight and it will be larger and very soft.

REASON: Most of the hard part of the eggshell is calcium carbonate. The acid of the vinegar attacks and combines with the calcium carbonate to form carbon dioxide gas which adheres to the egg in bubbles. The attached bubbles make the egg lighter, and allow it to float to the top. If it loses enough bubbles there, it sinks again.

By morning the hard part of the shell will have been dissolved, making the egg soft. By osmosis, some of the vinegar is taken into the egg through the softened shell, increasing the size of the egg.

PHYSICAL CHANGE OF A CHEMICAL

NEEDED: Sulfur in an old spoon and an alcohol lamp or hot plate with which to heat it.

EXPERIMENT: Heat the sulfur over the flame or the hotplate, and note the changes. It first melts into a watery, straw colored liquid. With more heat this shifts to orange and then to red. Then it becomes a slow flowing deep brown color, then almost solid, and finally becomes liquid again, and boils with a yellow vapor at 444 degrees centigrade.

REASON: The color of an element such as sulfur depends on its temperature and also on the size of the unit mass of the element and the number of atoms in the molecule.

SIMPLE CHEMICAL CHANGE

NEEDED: Sugar, an old spoon, and a stove.

SULPHUR

Physical Change
of a Chemical

COLD WATER

EXPERIMENT: Place sugar in the spoon, and heat it until the sugar has become a black charred mass.

REASON: Sugar is composed of carbon, oxygen, and hydrogen. The heat changes the sugar to carbon, which remains in the spoon, and water, which leaves the spoon as steam or vapor. The chemical change that has taken place is written: $C_{12}H_{22}O_{11}$.

CRYSTAL PALACES

NEEDED: Photographers' hypo, a pan, a stove, some water.

EXPERIMENT: Place the hypo in the pan, pour in a small amount of water, and heat on the stove. Stir constantly. When the hypo is dissolved, set the pan aside where it will not be disturbed.

As the water cools and evaporates, the hypo forms beautiful crystals, many of them in a form that suggests palaces.

Each salt has its own particular crystalline form. Try table salt in this experiment and see the difference.

SUPERSATURATION

NEEDED: Photographer's hypo (sodium thiosulfate), water, a measuring spoon, a small pan, a pane of glass or a mirror.

EXPERIMENT: Put five small spoonfuls (level full) of hypo into the pan, then add one spoonful of water. Heat gently until all the hypo is dissolved. Let the liquid cool almost to room temperature, then pour some out on the glass so that it covers the glass. Let it cool some more. Then drop one crystal of hypo into the center. Crystals will form into a beautiful pattern.

REASON: The liquid when cool is supersaturated, that is, it contains more of the dissolved hypo than it normally can hold. The forces in it are very delicately balanced, so that an addition of another crystal, or perhaps a jarring will start the crystallization process.

Cathodic Protection (1)

CATHODIC PROTECTION (1)

NEEDED: Three jars, three large nails, sandpaper or emery cloth, galvanized sheet metal from a tinsmith, tin can metal, water, salt.

EXPERIMENT: Clean the nails with sandpaper or emery cloth. Drive one through the sheet metal and one through the tin can metal. Place one in each jar (one nail without other metal, as a control) and cover with water. After several hours two of the nails will be rusted; the one in galvanized metal will resist rust, perhaps for days.

REASON: The zinc coating on the sheet metal, in electrical contact with the iron of the nail, protects the iron because it is more active and corrodes first. It is called a "sacrificial anode." The tin can metal, if enameled, has no such effect. If the tin can metal is coated with tin it actually speeds up the rusting of the nail.

The practice of using a sacrificial anode to protect the hulls of iron ships is called cathodic protection. Most of the sacrificial anodes for ships are aluminum. Magnesium rods are used for the purpose in water heaters.

Cathodic Protection (2)

CATHODIC PROTECTION (2)

NEEDED: Two nails, salt water, a flashlight cell, connecting wire.

EXPERIMENT: Connect the cell to the nails and place the nails in the salt water so they do not touch. One nail will rust quickly, the other will not.

REASON: One terminal of the cell produces an excess of electrons, and the wire carries them to one of the nails, giving it a minus (−) charge. That nail thus attracts hydrogen (+) from the water. The hydrogen coats the nail, protecting it from oxidation (rusting). The other wire from the cell draws some electrons from the other nail, giving it a positive (+) charge which attracts chlorine and other anions, forming oxidizing substances which attack the nail. Inside the cell one electrode is corroding and the other is not. The corroding electrode is connected by the wire to the protected nail.

This experiment illustrates control of the electrolytic corrosion of underground metal such as water pipes by the application of a negative electric current.

BLEACH

NEEDED: A glass half full of water, a few drops of red ink or food color, a teaspoonful of bleach.

EXPERIMENT: Mix the color and water, then add the bleach. The color disappears, leaving the water clear.

REASON: The bleach was made by reacting sodium hydroxide (lye) with chlorine to form unstable sodium hypochlorite, $NaClO$, which decomposes to liberate an abundance of nascent oxygen which is very reactive chemically. The oxygen atoms react with the color molecules to form colorless molecules.

Nascent, or "newborn" oxygen atoms are atoms that have not had time to combine into the oxygen molecule O_2. They are unattached, ready to attach themselves to other atoms or molecules. If they are combined into oxygen molecules they are not so reactive.

ANOTHER BLEACH

NEEDED: A glass jar, a small tin lid with a wire attached, some sulfur, matches, and a flower or apple peelings.

EXPERIMENT: Light the sulfur in the lid and let it down into the jar by means of the wire. Place the lid on the jar. After the flame has gone out, remove the dish of sulfur and place the flower or apple peels in the jar, covering it again quickly so that the gases do not escape. The flower or apple peelings will be seen to lose at least part of their color.

REASON: SO_2 (sulfur dioxide) gas formed by the burning S (sulfur) in the air will bleach many vegetable colors. It reacts with water to form sulfurous acid. The acid unites with the colored compounds in some organic substances and changes them to colorless compounds. Dried fruits have been home bleached and sterilized by this general process for many generations. A big tight barrel is generally used in "sulfuring" apples.

195

Another Bleach

Precipitation

PRECIPITATION

NEEDED: Epsom salts, ammonia solution, water.

EXPERIMENT: Make a solution of Epsom salts by dissolving some in water. Add a little ammonia solution. A white precipitate will be formed.

REASON: Epsom salts is magnesium sulfate. It reacts with the ammonia to form magnesium hydroxide, which is not soluble in water. It "precipitates" out in the form of white solid particles.

The chemical language for this is magnesium sulfate + ammonium hydroxide ammonium sulfate + magnesium hydroxide (precipitate).

EASY OXYGEN

NEEDED: A tall glass, cardboard, peroxide hair bleach from the drug store, baking soda.

EXPERIMENT: Put an inch of bleach into the glass, add a teaspoonful of baking soda, place the card over the glass to keep out the air, and let it sit an hour. Bubbles coming out of the mixture are oxygen. To prove it, place a glowing splinter of wood in the glass above the liquid and it will burst into flame.

REASON: The hydrogen peroxide molecules slowly change into plain water and nascent oxygen atoms. The O (oxygen) atoms unite into pairs to form oxygen molecules, O_2, which move into the airspace and accumulate.

Be sure the top of the glass is well covered by a smooth card so the oxygen will not escape. Put the splint into the airspace as quickly as possible after removing the card.

PAPER CHROMATOGRAPHY

NEEDED: An old dish, paper towel, ink.

EXPERIMENT: Place a piece of paper towel on the dish, put a drop of ink at a corner of it, and watch. The ink will be of varying colors as it is soaked up through the paper fibers.

REASON: Most inks are made up of different substances, and those travel through the paper at different speeds. Brown ink is very good for this; some inks may not separate at all. But this principle is dependable enough for its use in chemical analyses. It is called "paper chromatography."

Place a quarter teaspoonful of ink in the center of a paper towel, and as it travels through the paper fibers, place a half teaspoonful of water in the center of the circle.

INK ERADICATOR

NEEDED: Lemon juice and a household laundry bleach.

EXPERIMENT: Put a spoonful of water into a cup and add a spoonful of bleach. With a toothpick, place some juice on the ink letter to be eradicated. Let it stand a minute, blot it, then place bleach on it, again using a toothpick. The ink should disappear slowly. Blot again, and add a little more juice to prevent damage to the paper. Blot again.

REASON: Citric acid in the lemon juice reacts with a chemical in the bleach to form an unstable hypochlorous acid solution. This solution decomposes to liberate uncombined "nascent" oxygen atoms which unite with the dye that colors the ink. The dye molecules are destroyed, leaving a colorless residue.

SULFUR DIOXIDE

NEEDED: Small can, lemon, photographer's hypo, heat.

EXPERIMENT: Put a few crystals of hypo into the can with a little lemon juice and heat the mixture. The strong choking odor that comes off is sulfur dioxide.

REASON: The chemical action that takes place between the photographer's hypo and the citric acid in the juice is interesting. Hydrogen citrate in the lemon juice combines with sodium thiosulfate (hypo) to give thiosulfuric acid and sodium citrate. The thiosulfuric acid when heated changes to water, sulfur, and sulfur dioxide.

A "CARBON VOLCANO"

NEEDED: Sugar, baking soda, an old spoon or tin can lid, and heat.

EXPERIMENT: Mix thoroughly a teaspoon of sugar and a teaspoon of soda. Place them on the lid or old spoon and heat the mixture. The "volcano" will begin to erupt and the "lava" will rise.

REASON: The heat drives out some of the water from the sugar, leaving carbon. The heat also drives out carbon dioxide gas and water from the soda. The gas and steam make the carbon "rise" very much as carbon dioxide and water (steam) make bread rise.

One volume of carbon dioxide is produced to each 12 volumes of steam when heat decomposes the mixed solids, and the mixture becomes light and frothy. The black carbon makes the mixture black.

DISTILLED WOOD

NEEDED: A tin can with tightly fitting lid, a match, some wood shavings or sawdust, a nail and hammer, a heat source.

EXPERIMENT: Make a small hole in the lid of the can with the nail. Place the wood in the can, cover with the lid, and heat the can on the stove. Steam will come from the hole, then gradually a gas will begin to come out. It will be a different color. The gas may be lighted with a match and it will burn. When no more gas comes out, remove the can from the heat, let it cool, and the wood will be seen to have turned to charcoal.

REASON: Wood is made of cellulose, with minor impurities of several kinds, all of which are mostly carbon, hydrogen, and oxygen. As the temperature goes above the boiling point of water, the water is driven off as steam. Soon tars and gases begin to come out of the hole and these will burn. Charcoal, which is mostly carbon, is left. This is the way artificial cooking gas is made, except that coal is used instead of wood, and the tars are saved and converted into many useful chemicals.

CHEMISTRY IN THE TOASTER

NEEDED: White bread toast, household iodine, water, a glass and a dish.

EXPERIMENT: Mix a teaspoonful of iodine in half a glass of water, and pour it into the dish. Dip the edge of a strip of toast into the solution. The center will turn bluish purple while the toasted part remains unchanged.

REASON: Iodine is a test for starch. The untoasted part of the bread turns purple, showing it is mostly starch; the toasted part remains unchanged because its starch has been changed by heat into dextrin. Dextrin iodide is not bluish purple. Starch iodide is.

Toasting is a step toward digestion of the bread. This is why toast rather than plain bread is recommended for sick people.

A CHEMICAL TEST

NEEDED: Boric acid, denatured alcohol, food jar lid, match, stainless steel spoon.

EXPERIMENT: Put a few crystals of boric acid into the lid. Add a teaspoonful of denatured alcohol and stir with the handle of the stainless steel spoon. Light the mixture and the flame will be seen to have green edges. As the alcohol burns away the flame will be a very beautiful green color. This experiment should not be performed in bright light.

REASON: Boron in the boric acid, H_3Bo_3, is heated to incandescence by the heat as the alcohol burns. The H part of the burning acid gives an unnoticed pale blue. The boron gives a noticeable bright green flame. Other elements give different colors in the "flame test."

The Flame Test

THE FLAME TEST

NEEDED: An alcohol lamp or Bunsen burner, table salt, baking soda, small pliers, steel wool, a small copper wire.

EXPERIMENT: Hold the end of the copper wire in the flame; note the green color given to the flame. Hold a small piece of steel wool in the flame with the pliers and notice the color. Drop a little of the salt and soda into the flame and notice the color.

REASON: Each vaporized metal has its characteristic flame color, and many can be identified in this manner. The salt and soda show the same color because it comes from the sodium, which is one of the elements in each. Some single colors are so intense that they hide other colors which may be present. Certain types of compounds vaporize more readily than others in the same flame.

SOLVENTS

NEEDED: Carbonated soft drink, salt, a dime.

EXPERIMENT: Make sure the mouth of the bottle is wet. Drop a little salt into the drink. Place the dime on the mouth of the bottle. It will move up and down, showing that gas is coming out of the bottle.

REASON: The water of the soft drink has carbon dioxide, a gas, dissolved in it. Salt water can hold less carbon dioxide in solution than plain water. When the added salt dissolves in the water some of the gas has to come out.

This does not mean that water cannot hold many different substances dissolved in it. It can. In the soft drink there are many dissolved substances in addition to the salt and the carbon dioxide. It just cannot hold as much carbon dioxide when salt is added.

Incidentally, if there are possible harmful chemicals in the drink, carbon dioxide is *not* one of them.

CANDLE FLAME CHEMISTRY

NEEDED: A burning candle, a medicine dropper, some matches.

EXPERIMENT: Remove the rubber from the dropper, and hold the large end of the glass in the candle flame. Keep trying, and you will find a way to hold it so that gases coming through the dropper may be lighted at the little end. (Careful—the glass gets hot.)

REASON: The liquid paraffin climbs up the wick by capillary action, and is heated there and turned into gases. Ordinarily these would be burned in the outer portions of the flame, but they may go up the cool glass tube and be lighted at the upper end.

Much complicated chemistry goes on in the burning candle. Some of the substances found there are vaporized paraffin, ethylene, carbon dioxide, carbon monoxide, water vapor, hydrogen, oxygen, and nitrogen (from the air). The end products of the burning are mostly carbon dioxide and water.

LEAD CHEMISTRY

NEEDED: Inexpensive lead acetate from the drug store, zinc from a flashlight battery, distilled water, jars, soap and water, steel wool.

EXPERIMENT: Stir half an ounce of lead acetate in a pint of distilled water. Some tap water might do, if it is comparatively free of minerals. Let the solution sit overnight, and pour off the clear liquid for use in the experiment.

Scrub the zinc with soap and water, then with steel wool until it is bright. Suspend the zinc in the solution. Immediately it begins to turn black, as zinc replaces the lead in the solution. The lead deposit continues to grow on the zinc, and glistening spots may be seen. Leave it overnight. The zinc will be practically gone by next day. Take it out of the solution, let it dry a day or two, and it will be covered with bright shining lead crystals.

REASON: Zinc is more active than lead, which means that zinc atoms readily lose electrons if some units which absorb electrons are at hand. Lead in the acetate solution is in the form of lead ions which can attract and hold two electrons per ion. Zinc atoms lose two electrons per atom and become soluble zinc ions. Lead ions become insoluble lead atoms which unite to form glistening crystals.

Note: Lead acetate (common name sugar of lead) is poisonous if swallowed.

MORTAR AND CONCRETE

NEEDED: Lime, cement, sand, water. A builder will probably give a little lime, cement, and sand from a building site.

EXPERIMENT: Wash the sand in a jar or can to remove the dirt. Mix one spoonful of lime with water so that it makes a paste. Put in four spoonfuls of sand and mix thoroughly. Let it dry slowly on a metal lid. This is mortar.

Mix three spoonfuls of sand and one of cement with enough water to make a paste, and let it dry slowly. This is concrete. The drying should take place under a moist cloth to prevent too-rapid a loss of water.

REASON: As water evaporates from the mortar mix, carbon dioxide from the air takes its place, forming calcium carbonate, which is the chief part of natural limestone. Concrete dries by combining with water to form a hard firm complex. It must dry slowly, and it gets harder as it ages. Several years may elapse before it reaches maximum hardness.

The words "cement" and "concrete" are used sometimes to mean the same thing: the finished concrete. This is incorrect. "Cement" is highly active powder before it is mixed with sand and water. It is made by burning together limestone and clay or shale, then powdering the clinker. Just how it hardens into the popular building material so widely used is not adequately explained.

SUBSTANCES THAT SUBLIME

NEEDED: A jar with a loosely-fitting lid, a pan, some mothballs, water, a stove.

EXPERIMENT: Place the mothballs in the jar, cover with the lid (do not tighten the lid!), place the jar in the water in the pan, and heat the water until it is ready to boil. Turn off the heat and watch. Notice that the mothballs do not melt and become a liquid, but turn directly into a gas. They "sublime." Some of the gas condenses on the sides of the jar, making crystals. If the gas is allowed to flow out of the jar the mothballs eventually disappear, more rapidly if they are again heated.

REASON: Most substances melt and pass through the liquid state before changing into a gas. Water does usually, but clothes on a line will "freeze-dry"—that is, will dry when the weather is too cold to let the water in the wet clothes become a liquid.

CHANGING CHOCOLATE

NEEDED: Two chocolate bars.

EXPERIMENT: Let one of the bars get warm in the sun or elsewhere. Allow it to cool again. It will have become yellowish-white in color. Compare with the other.

REASON: According to Hershey's L.F. Santangelo, chocolate softens at 80 degrees, and melts at 92 degrees. At this temperature some of the cocoa butter separates from the other ingredients, then when the bar hardens again, the cocoa butter crystallizes on the surface as yellowish white crystals. This perhaps makes the chocolate less appetizing, but does not affect the nutritional value or the taste.

THE KITCHEN MATCH

NEEDED: A kitchen match.

EXPERIMENT: Strike the match. Note carefully how it burns when held horizontally.

REASON: When the head is rubbed a little heat is produced. Not much, but enough to light the tip, which is probably phosphorus sulphide and potassium chlorate and a binder. The burning tip lights the bulk of the head of the match, which is probably potassium nitrate and charcoal with a binder. The burning head lights paraffin in which the end of the match stick has been dipped, then the burning paraffin lights the wood which is probably pine.

The other end of the stick (that end held in the fingers) had been moderately "fireproofed" by dipping in a chemical (probably ammonium phosphate) intended to actually retard the flame, to prevent the stick from glowing when discarded.

THE SAFETY MATCH

NEEDED: A match book.

The Safety Match

EXPERIMENT: Strike a safety match on the part of the book intended for the purpose. Also, rub the match quickly over a glass surface and see if it will strike.

REASON: The head of the safety match is composed of antimony sulfide and potassium chlorate, with a binder to hold them together. This mixture does not ignite easily by the heat of friction, but is lighted by any tiny flame.

The "strikum" on the paper book is composed of red phosphorus, powdered glass, and glue—the binder. As the head of the match is rubbed across the "strikum," a little of the phosphorus is torn off and takes fire in contact with the very combustible head of the match. The head takes fire and lights the paraffin soaked stick. A safety match may be struck on glass because the smooth surface can allow a build-up of heat by friction—enough to fire off the head before the head wears out, as it would do on most surfaces.

Match companies do not all use the same ingredients in their matches. Some other chemicals used are red lead, lead dioxide, and manganese dioxide. White phosphorus is no longer used because it takes fire too easily and is poisonous.

There are some safety matches difficult, if not impossible, to strike on glass, because of the difference in ingredients used. Also, it takes a little practice. Motion must be rapid and pressure just right.

Chapter 14

Mechanics

THE BRIDGE

NEEDED: Regular cardboard, corrugated cardboard, books, weights.

EXPERIMENT: Place a strip of regular card about 2 × 15 inches across books to make a bridge. It will hold up little or no weight. Place a similar strip of corrugated board across, and it will

The Bridge

hold up a weight such as a glass paper weight. Split the corrugated card, along its length as shown, and it will be weak, like the regular card.

REASON: The corrugations between the flat card form triangles that are very strong in resisting compressional and tensile forces. In other words, to bend the card the corrugations would have to be pushed or pulled with enough force to deform them.

When the card is split through the corrugations the sections of it will flex, or bend when only a small force is applied. There is practically no compression and tension to be resisted. The split card behaves like the regular card.

Friction

FRICTION

NEEDED: Roller skates.

EXPERIMENT: Have the skates pointing forward, and try to walk. Moving forward is difficult because the skates offer little friction between the feet and the floor. This is rolling friction, and since the wheels are between the feet and the floor it is easy for the wheel to roll but not easy to move the body forward.

Without skates the shoes against the floor offer so much friction that they slide only with some difficulty.

COMMENT: Friction can produce heat and wear in bearings and be very destructive, but it is useful in almost all our actions.

Even in sleep it is useful; it keeps us from sliding off the bed.

Without friction an automobile could not move and without friction it could not stop once it started moving.

The Compound Pendulum

THE COMPOUND PENDULUM

NEEDED: A paper cone (a drinking cup will do), two lengths of string, a doorway or other support, a large square of cardboard, preferably dark in color, some salt.

EXPERIMENT: Hang the strings from two places at the top of the doorway, tie the cup to each string as shown in the drawing. Make a hole 1/8-inch in diameter in the bottom of the cup.

Draw the cup to one corner of the cardboard, fill it almost full of dry salt and release it. As the cup swings this "compound pendulum" will make various and interesting designs with the salt. Change the length of the upper string or the lower string, try again and the design will be different.

REASON: The strings and cup act as two different pendulums swinging back and forth through the doorway, also to and fro across the face of the doorway. This combined motion causes the cup to swing in a regular pattern giving the beautiful designs.

ANGULAR MOMENTUM

NEEDED: An object on the end of a string.

EXPERIMENT: Whirl the object around on the string. Extend a finger so that the string begins to wind up on it. The finger may be held still, yet the object will continue to whirl, faster and faster as the string is wound up on the finger, until finally it touches the finger.

REASON: The angular momentum is the mass of the object, times the angular velocity (that is, the rate of spin), times the length of the string squared. The angular velocity remains constant. Since the mass of the object is constant, the rate of spin must increase as the string gets shorter to keep the product unchanging. So, the object spins faster and faster.

Fluidics

FLUIDICS

NEEDED: A steady flow of water from the tap, a soda straw, a glass.

EXPERIMENT: Fill the mouth with water, then blow the water out of the mouth through the straw so that it hits the larger stream from the tap. The smaller stream from the straw will turn the larger stream aside.

REASON: This is the principle of the fluid amplifier device in a new technology called "fluidics." The weaker stream of water changes the larger stream, much as a weaker current or charge can change a larger current in electronics. Fluidic devices are replacing electronic devices in many control systems today. They have advantages in some cases.

HAMMER—HEAD

NEEDED: A nail, a hammer, a plank, a large book such as a heavy dictionary.

EXPERIMENT: Place the book on someone's head, place the plank on the book, and the nail may be driven into the board without hurting the head.

REASON: There is much unfilled space between each page of the book and this space contains air which acts as a cushion. The mass and weight of the book absorb much of the blow of the hammer, because of their inertia. The book also spreads the force of the blow over a wide area of the head. If the book used is small, a tack hammer and a tack should be used instead of a hammer and nail.

Stick-Slip Friction

STICK-SLIP FRICTION

NEEDED: A violin or a rosined string.

EXPERIMENT: Draw the bow along the violin string, and notice that the string vibrates, or moves rapidly back and forth.

REASON: As the bow moves, the string sticks to it and moves with it. Then the string slips and flies back. The bow sticks to it again, and moves it again. The process is repeated to make the musical note.

This type of friction may be shown by drawing a rosined string along the edge of a tin can or through a hole in the bottom of a can, or the string can be attached to the bottom of a can, stretched, and rubbed with a pencil. An unpleasant growling sound is made in this way, whereas a musical note is produced by the tuned violin string. Slip-stick friction may also be shown by rubbing the finger across a table as shown in the upper drawing.

RUBBER BANDS

LARGE NUTS

The Homing Can

THE HOMING CAN

NEEDED: A coffee can, rubber bands, a heavy iron bolt or nuts.

EXPERIMENT: Fasten the bands to the bottom and lid of the can, and attach the iron object in the middle of the bands. Put the lid on. Roll the can away and it will return.

REASON: As the can rolls away the hanging weight hangs and does not turn as the bands twist. This twisting builds up as energy in the bands. As the can stops rolling the energy can be released from the bands as they begin to untwist, rolling the can along in the opposite direction.

THE MYSTERY SPOOL

NEEDED: A spool with thread wound around it.

EXPERIMENT: Place the spool on a level surface. Pull the thread as if to unwind it, and the spool will roll toward you winding up more thread.

REASON: The frictional force acting where the spool makes contact with the surface causes a greater torque about the center of the spool than does the force due to the pull on the string. The result is angular motion in a direction to wind up the string.

Newton's Third Law

NEWTON'S THIRD LAW

NEEDED: A smooth board and sticks to hold a pendulum, large soda straws, string, large bolt or other weight.

EXPERIMENT: Build the apparatus as shown, so the board can move on the straws. As the pendulum swings the platform moves back and forth in the opposite direction.

REASON: Newton's third law states that for every action there is an opposite and equal reaction. At the end of a swing the string pulls one way on the weight and the other way on the platform, and so both move.

When the support is allowed to move in this way the pendulum behaves as if it were shorter than it really is. The "virtual pivot" is between the support and the bob (weight).

The I-Beam Idea

THE I-BEAM IDEA

NEEDED: A yardstick and a weight.

EXPERIMENT: Hold the yardstick by the ends, with the weight suspended from the center. If the stick is on edge, it does not seem to bend. If it is flat, it bends under the weight.

REASON: Engineers have discovered that a support beam varies in stiffness as the cube of the vertical height, other factors being equal. In the upper picture the cube of the vertical height is many times greater than the cube of the vertical height in the lower picture. Steel I-beams have much greater stiffness if the I is vertical. Observe the construction arrangement of the steel frame of a new building. Fishing rods bend easily at the tip where the vertical thickness is very small.

PULLEYS

NEEDED: Pulleys, cord, a support, weights.

EXPERIMENT: String the pulleys in various ways, and see how the weight is easier to move as more strands of cord are used.

REASON: The actual weight of the object is felt as the cord is pulled downward over a single pulley. If the pulley is placed at the bottom and the cord pulled upward, the weight seems to be about

Pulleys

half as much, but the cord has to move twice as far as the weight moves.

A block and tackle, which consists of several pulleys and several strands of rope increases mechanical advantage. But as the number of strands supporting the weight is increased the applied force must act over a longer distance. In the picture the second pulley system has the greatest mechanical advantage where four strands support the weight. The third and last arrangements have the same mechanical advantage, since each has two cords supporting the weight. In any case if the pull is vertical the theoretical mechanical advantage equals the number of strands which support the weight. More work must be put into any machine than is gotten out of the machine. The extra work must be done because of friction.

FOUCAULT'S PENDULUM (1)

NEEDED: Heavy weight, strong string, a solid support more than ten feet high, a four-foot stick.

EXPERIMENT: Start the weight swinging in a straight path, and place the stick under that path, to mark it. After ten or more minutes the weight will have changed its path of swing slightly—a proof of the turning of the earth.

REASON: Jean Bernard Foucault was a French physicist (1819-1868). He made discoveries in light, electricity, and magnetism, and invented the gyroscope. His pendulum experiment, proving that the earth turns, was made in 1851.

This is not an easy experiment to do at home. For a weight, the author used a 21-pound iron ball. A water bucket full of wet sand might be used. The author attached a bracket to a large tree limb 19 feet above the ground, and allowed the nylon cord to swing from it. A breeze or the untwisting of the cord or possibly other factors may interfere.

Foucault's Pendulum (2)

FOUCAULT'S PENDULUM (2)

NEEDED: Phonograph, large glass jar, stick, string, weight.

EXPERIMENT: Set up the apparatus as shown. Start the weight swinging, then start the turntable at its slowest speed. Note that the plane of swing of the weight does not change as the jar turns.

RESULT: The Foucault pendulum continues to swing in the same plane as the earth turns under it. If a phonograph is not at hand, a revolving stool may be used. Be sure the weight (which can be a fishing sinker) is suspended from the center, otherwise centrifugal force as the turntable turns may throw the experiment out of balance.

Tricky Marbles

TRICKY MARBLES

NEEDED: A ruler with a groove along its length, a dish, two marbles.

EXPERIMENT: Place a marble on the dish, hold the other in the groove on the ruler and aim the groove at the first marble. When the second marble is allowed to roll down and strike the first, it stops and the first marble is catapulted out of the dish.

REASON: The marbles are elastic. When one hits the other, its force is transmitted into the second. Since the collision of the marbles is almost perfectly elastic, the momentum of the first marble is transferred to the second marble. The first marble stops and the second marble moves off with the original velocity of the first.. (Glass is more elastic than rubber).

WHICH FALLS FIRST?

NEEDED: A yardstick or meter stick, a coin such as a quarter or half dollar.

EXPERIMENT: Place the stick against the wall so that it will not slip. Place the coin on the end, as in the upper drawing, and when the stick is released the coin will not fall as rapidly as the stick. Place the coin back one-third the length of the stick, and both the coin and stick will fall together.

Which Falls First?

CENTER OF GRAVITY

REASON: The stick is not falling as freely as the coin. The stick is rotating about an axis at the wall. The velocity of the end of the stick is greater than the falling coin. The velocity of a point on the stick one-third the length of the stick from the end moves at the same velocity as a freely falling body.

THE TOUGH BALLOON

NEEDED: A balloon.

EXPERIMENT: Blow air into the balloon, lay it on the table, then burst it by placing a smooth surface, such as a book, on it and pressing down. The effort required to burst the balloon will be surprising.

REASON: Pressure is force per unit area. Here the pressure difference on the inside and outside of a toy balloon necessary to burst it is not very great. By putting the book on top of the balloon the force (push) is distributed over such a large area of the balloon that the pressure increase is small. A small push on a pin would cause a large pressure which would easily break the balloon. Total force = pressure × surface area.

A BAD BALANCE

NEEDED: Ruler and string, coins.

A Bad Balance

EXPERIMENT: Suspend the ruler on the string above a table top, and try to balance coins on it. It does not work. Put a screw eye in it, tie the string to it, and the ruler becomes a balance that can be used to show the principle of the lever.

REASON: The ruler cannot balance if suspended by a string from the lower side, because the center of gravity is above the suspension point or fulcrum. Either one side or the other will go down and touch the table top. Suspending it by the screw eye places the balance point above the ruler, and a balance can be effected as long as the coins are not stacked too high.

METEORS IN THE HAND

NEEDED: Flint rock and a piece of steel, or a cigarette lighter.

EXPERIMENT: Strike the rock with the steel, or turn the wheel of the lighter. Sparks will fly.

REASON: When the steel is rubbed over the flint quickly, some tiny pieces of the flint are rubbed off. The combination of the friction and the high speed at which they leave the flint heats them to the burning temperature of dry tinder or lighter fluid.

This is something like the meteors in the sky. Their speed

through the air causes friction which heats them to a high temperature so that they give off light.

CRACK OF THE WHIP

NEEDED: A whip that will "crack."

EXPERIMENT: Try to think of the explanation of the loud sound made by the whip as the wave goes to the end, faster and faster, finally ending with the "crack."

REASON: It could be called an example of conservation of angular momentum; the momentum starting at the thick part of the whip continues, but the smaller parts of the whip must move faster to maintain the momentum.

The end of the whip moves faster than the speed of sound; it "breaks the sound barrier," moving faster than a .45 pistol bullet! The whip does not slap back on itself to make the noise.

Uphill Roll

UPHILL ROLL

NEEDED: Two conical paper cups, two yardsticks, three books, gummed tape or needle and thread.

EXPERIMENT: Sew or tape the cups together so that they make double cones. Arrange the yardsticks so they are held by the book in an upward slanting position as shown. The cups, placed at the bottom, roll to the top of the yardstick incline.

REASON: The cups do not actually roll uphill. Because of their shape, their center of gravity moves down as they roll to the upper ends of the spreading sticks. For proof, notice how far they are from the table at the beginning of their journey and at the end. They are closer to the table at the end.

The Rolling Marble

THE ROLLING MARBLE

NEEDED: A ruler, a thick book, a marble, a rug on the floor.

EXPERIMENT: Place the end of the ruler on the edge of the book, let the marble roll down the ruler and measure the distance it rolls. Place the ruler so that the book is under the middle of it, let the marble roll from this middle point, and measure the distance it rolls. The distances rolled should be the same.

REASON: The original potential energy of the marble is the same in both positions mentioned above. Potential energy is the product of weight times height. This energy is converted to kinetic energy of motion in rolling down the ruler, therefore the speed of the marble is the same when it leaves the ruler in both cases and should roll the same distance along the rug. A very soft or deep pile rug is not too satisfactory for this experiment.

More Rolling

MORE ROLLING

NEEDED: An incline, a roller skate, a ball, a cylinder, and a disc or wheel.

EXPERIMENT: Release all the rolling objects at the same time at the top of the incline, and see which reaches the bottom first.

REASON: All the objects have the same potential energy per pound at the start. As they move down the incline their potential energy is changed into kinetic energy, energy of motion, which is in two forms, translational and rotational. The translational energy is what gives a body a velocity straight ahead, while the rotational energy is tied up in making the mass of the body turn or spin. The skate has the greatest percentage of its energy as translational kinetic energy, the least as rotational kinetic energy, and so has a higher velocity all the way down.

The incline must be smooth, so that the effects of rolling friction need not be taken into consideration. See which object rolls farthest after leaving the incline.

THE SWING

NEEDED: A rope swing.

EXPERIMENT: Twist the rope, and let it start to unwind with your legs outstretched; the spin will be rather slow. Draw in your legs, and the spin will be faster. Extend your legs, and the spin will slow down again.

REASON: This is another study in conservation of angular momentum, which, for each particle in your body, is its mass times its angular velocity (rate of spin) times the square of its distance from the center about which you are turning. If you change the distance of some parts of your body from the center, the spin will have to adjust itself so that the angular momentum stays the same, (stick legs out and slow down, pull them in and speed up).

Skaters and ballet dancers use this principle. They start a spin or whirl with arms extended, then spin faster as they draw their arms in closer to their bodies.

Index

Index

227